BOOKS BY HARRY HOPKINS

THE NUMBERS GAME

THE NUMBERS GAME

The Bland Totalitarianism

by Harry Hopkins

Little, Brown and Company — Boston — Toronto

FIRST AMERICAN EDITION

Library of Congress Cataloging in Publication Data

Hopkins, Harry.
   The numbers game.

   1. Statistics. I. Title.
HA29.H73   301.2'1   73-4002
ISBN 0-316-37270-6

The tendency to fatalism is never far from mankind. It is one of the first solutions of the riddle of the earth propounded by metaphysics. It is one of the last propounded by science.... No race is naturally less disposed to a fatalistic view of things than is the Anglo-American.... Nevertheless, even in this people, the conditions of life and politics have bred a sentiment or tendency which seems best described by the name of fatalism.

James Bryce, *The American Commonwealth*, Chapter XXXIV, 'The Fatalism of the Multitude'.

There is more than a germ of truth in the suggestion that, in a society where statisticians thrive, liberty and individuality are likely to be emasculated.

M. J. Moroney, *Facts from Figures*

WHAT IS THE NUMBERS GAME?

Why is it that with the cornucopia of 'the consumer society' spilling out before us we so often feel that we 'have no choice'? The key to the paradox lies in the Numbers Game, that central phenomenon of our times by which our lives are dominated by statistical artefacts, from the IQ score which switches a child into its life-track to the Growth Rate which shapes whole societies. Replacing 'the stars in their courses' which guided the destinies of a simpler age, the dominion of the Indices and Indicators has stolen up on us, in the last fifty years, through the convergence and interaction of a number of factors—the rise of statistical and social science, the mushrooming of the mass media, the sophistication of marketing, the advent, most notably in America, of fully fledged electronic mass democracy. The result has been the complex, semi-automated system of feedback here called the Numbers Game. In essence, the mechanism is that the dehydrating statistical processing favoured by the social sciences, particularly by economics and experimental psychology, create a value-vacuum—or a continuous suspension of judgment—in which the associated statistical processes of marketing and media take over. The resultant 'reality' is then measured and validated by the statistical accountancy of social science, authenticated with the stamp of Democracy, and lent further authority by the ancient magic of number.

In this and other complicated ways the Numbers Game is both self-feeding and self-perpetuating.

CONTENTS

I

THE HYPOCHONDRIACS

... it is lack of confidence more than anything else which kills a civilisation.

Sir Kenneth Clark, BBC series 'Civilisation'.

'Never has a nation seemed to have had more, and enjoyed it less,' declared President Nixon in his 1970 State of the Union message. 'The glitter of growth,' admitted *Time* magazine, 'has begun to tarnish.' To a European at least, Americans seemed uneasily poised just then between novel misgivings about the old ideal of endless economic expansion—'the Permanent Revolution' so confidently hailed by the Editors of *Fortune* as recently as 1951—and older, sharper, recurrent fears that somehow, one day, that splendid American process might finally falter. In the summer of 1970 a survey carried out in key areas of the United States by the *Washington Post* confirmed that most of the citizens of the world's most affluent nation were worried and fearful about the future.

Whether the recession of 1969/70 really was a 'recession'—or, as Dr F. Thomas Juster of the National Bureau of Economic Research later suggested, merely 'a trend'—there seems little doubt that it left its mark. For, however called, it was the fifth such bout of economic heavy weather since the war, and for months, from the spring of 1969, the economic augurs had been huddled over their econometric models, scrutinising each new economic 'indicator' as it emerged and issuing cabalistic, increasingly ominous, warnings. In April 1969, in order to 'restore stability and reduce inflationary pressures', the Federal Reserve Bank had raised the Discount Rate to 6 per cent—the highest for forty years. But inflation continued at 6 per cent, reaching its highest rate since the Korean War in 1970. Wall Street turned steeply downwards. Unemployment soared. The ghost of 1929 walked again, although some economists expressed a preference for the ghost of 1907, a liquidity panic.

In August 1969 the Washington economic columnist, Eliot

Janeway, wrote: 'The economy is running like a jet plane which is burning up its last reserve tank of fuel ...'[1] In October, with all eyes fixed on the altimeter of Wall Street, the key level was said to be 800: if the Dow Jones Industrial Average went through that, a further plunge, and possibly a deep recession and mass unemployment, might lie ahead. In November the Dow Jones did just that. In the last quarter of the year there was a minute decline in the Gross National Product; and according to the Department of Commerce definition a 'recession' exists when output in real terms shows no growth for two consecutive quarters. In the first quarter of 1970, the GNP fell again. It was noted that this was its biggest fall for over ten years.

The critical numbers mounted up. Unemployment rose to almost 5 per cent, around 4½ million people workless. Wall Street's plunge deepened: on 25 May the Dow Jones Industrial Average sagged to 641, headlined as its 'Biggest Fall in 6½ Years'. According to Mr Monte Gordon, a vice-president of Balche and Company, the dive represented 'the crushing impact of disappointed hopes'. Stock markets almost everywhere in the world reeled under the impact of Americans' crushed hopes.

In July, the US Department of Commerce noted that the Gross National Product had grown by 0·3 per cent in the second quarter. Hope, it seemed, was reborn. The 'jet plane' pulled out of its dive.

Yet, as the metaphors suggest, the cumulative psychological cost of maintaining the equilibrium of a capitalist economy is not small. 'In the American economy, if you do not go up, you go down,' explained Dr Arthur M. Okun, 1968 Chairman of the Council of Economic Advisers. 'The potentiality for a stall is a real danger.'[2] By 1970 the American public needed little persuading of this fact. According to a Louis Harris poll taken in mid-August 1970, 58 per cent of Americans expected the recession—which they had no hesitation in calling a recession —to continue; and 44 per cent expected that prices would rise faster than incomes for the next few years. That is, Americans would get poorer.

In their economic sense the words 'boom' and 'slump' are of

American provenance, and when they first appeared in the London *Daily Mail* around the turn of the century, the proprietor, Alfred Harmsworth, ordered that they should do so in inverted commas. Today, however, there is nothing peculiarly American about this co-existence of anxiety neuroses and affluence. It is 'the way we live now'. With some changes of accent and emphasis, President Nixon's rueful remarks about America could equally apply to postwar Britain, where two decades of unprecedentedly widespread prosperity had been cankered by obsessive anxieties about both present and future, and despite a Prime Minister's celebrated reminder to fellow citizens that they had 'never had it so good', many were morosely convinced that things had never been so bad, and were getting steadily worse.

It was a paradox which declared itself in almost every week's newspapers in these years, and one which may well puzzle historians in the future. For by any reasonable standard Britain's achievement since 1945 had been far from justifying any failure of confidence. Despite the strains of five years of total war, the liquidation of her large overseas investments, and, now, the military support and subsidisation of her chief competitor, Germany, Britain was still the world's third largest exporter, and, indeed, if one included services as well as merchandise, she ranked first in the world in terms of export earnings per head.[3] Only one other major country, West Germany, exported so much—around a third—of her engineering output. And in a dozen directions since the war British scientists and technologists had shown that the inventive genius which had done much to make Britain the world's first industrial nation had by no means deserted her.

Public service expansion—31 new towns, 29 new universities, a comprehensive national health service—had been matched by private gains. In the sixties alone the average British standard of living, in real terms, measured by consumer expenditure, had risen again by a third; the number of private cars had almost doubled; supermarkets had multiplied by ten. Most significant, perhaps, was the explosion of what were now being called 'the leisure industries': those taking holidays abroad had grown from one million in 1958 to five millions in 1965.

And yet—in President Nixon's words—never had a nation seemed to have had more, and to have enjoyed it less. The same newspapers which recorded the statistics of affluence contained column after column of self-denigration and 'un-English' self-doubt. And somehow the statistics on which headline attention was focused were not the statistics of solid achievement, but the statistics of nervous debility and perpetual crisis. Like a child seen through the eyes of an over-anxious mother, the British economy was ever in a delicate state. If it was not being sapped by something called 'slow growth' it was menaced by 'overheating'. On occasion it seemed afflicted by both at the same time. Figures in the headlines kept the nation swinging between flickering hope and rising despair. The Gold and Dollar Reserves dwindled. The GNP rose slightly, then sank back exhausted. The crime rate soared.

And while it was true that the Pound Sterling, as one of the world's two major reserve and trading currencies, was inevitably vulnerable in the post-Imperial world (with Commonwealth war credits and Sterling Balances piled up in London), there was surely something almost masochistic about the way in which the British themselves turned the spotlight on the sterling exchange rate as 'Britain's Virility Symbol', inviting the world, so to speak, publicly to establish her impotence. Certainly, had the Victorians—no strangers to massive debts and financial panics—been able to return to this strangely unconfident Britain they would have found it singular beyond words that their descendants should have chosen to centre so much of their lives and hopes, year after year, around the *monthly* figures for the *visible* Balance of Trade—that favourite postwar British exercise in self-defeat, 'the Trade Gap'.

For the 'trade gap' or deficit on 'visible' or merchandise trade had indeed existed for over a century, through most of the years when Britain was the richest country in the world. It had normally been covered by the surplus on the 'invisibles', overseas investment income, insurance, merchanting, banking, shipping and so on. And if the 'invisibles' now paid a smaller proportion of the import bill than in earlier days, the significant thing was that, by and large, taking *not*—as was patently absurd—each month, but one year with another, the shortfall had been filled.* Even if there were to be a deficit on a year's final Balance of Trade, in the sixties it was unlikely to amount to more than 1 to 1½ per cent of total national output,[4] and could be fairly readily set right in a number of simple enough ways if need be. Certainly, there was no cause here for panic and despair. Yet panic and despair were the British public's almost monthly lot, as, with the notable aid of press and Board of Trade—which persisted in publishing the monthly 'Gap' figures without the compensating 'invisibles' (only available at a later date)—it fixed its eyes on the figures of Doom ('roaring ahead', 'leaping', 'widening', 'sliding') and resignedly gave itself up for lost, like a suicide rooted in the path of an oncoming train.†

By the spring of 1971 when Britain had, in fact, achieved a record balance of payments surplus—now being described as 'embarrassingly large'—the London *Observer* greeted the year's Budget and the daffodils with a series of top-of-the-page charts showing the GNP *Stagnating*, Investment *Flagging*, Prices *Rocketing* and Unemployment *Soaring*. 'Next year,' added its resident Professor of Economics, striking the inevitable note with unerring precision, 'the Government may not be so lucky ... the increasingly practical question is not whether we shall get another payments crisis, but when. Will it be 1972—or can it be held off till 1973?'

Was consumer expenditure now rising? Britain was confronted with the spectre of a mounting inflationary spiral. But consumer expenditure was flagging. Worse! Britain would perish from those ever-threatening maladies, constricted investment and slow growth. After two decades of such incessant nagging by omniscient economists and battering by the indices, few could be surprised that a national poll showed that two in every five Britons would like to move elsewhere. If challenged, many conscientious daily meter-readers would now

---

* According to Mr Harold Lever, Financial Secretary to the Treasury until 1970, over the years between 1951 and 1968, taking one year with another, the Balance of Trade had in fact yielded a modest surplus on current account. No one, however, would ever have thought so from the beating of breasts which went on on this subject.

† Not until the summer of 1970 was the form of presentation of the Balance of Payments statistics changed, and, so to speak, by the stroke of a pen, the earlier yawning chasms belatedly bridged.

judiciously explain that the nation was in an—apparently ir-reversible—moral decline, work-shy, stagnant, class-bound, and generally in the category that Australians call the 'no-hopers'.

The Metered Life

What had become of that famous British phlegm which had enabled the country to stand alone in 1941 against Hitler as earlier it had stood alone against the all-conquering Bonaparte? Where was that robust commonsense which had made British political stability the envy of Europe?

It seemed to have gone the way of the old American optimism, the invincible conviction that 'the impossible will take a little longer' which as recently as 1945 had still enabled the Americans to see 'the American Way of Life' as sovereign and sufficient remedy for all the world's ills. (Who would have dreamed that by 1970 *Time* magazine would be sourly report-ing that ' "America the Efficient" seems to have become a land governed by Murphy's Law: if anything can go wrong, it will —and at the worst possible time.')

Indeed, both American and British national characters, which in their very different ways had done so much to form the mod-ern world, appeared to be in process of deep transformation. What lay behind this critical, and possibly secular, change?

Britain, certainly, had not long since lost an Empire; and the United States, since Sputnik I shot across the skies in 1957, had lost its once irresistible monopoly of 'know-how'. For the British there had been the trauma of Suez; for America the running sore of Vietnam. These were severe shocks; but the British had lost empires before, and America remained beyond compare the industrial colossus of the world.

There were many other factors, some world-wide; the up-surge of the non-white peoples, the violent rejection by the affluent young of the world of their fathers, the loss of the assur-ances of religion. All these and many other causes of the current loss of confidence have been persistently examined. Yet there is one factor which has largely escaped scrutiny, possibly because its very quiet pervasiveness renders it invisible. This forgotten factor is the proliferation and institutionalisation of the social sciences and, accompanying this from 1940 onwards, what one authority calls 'a phenomonal increase in the number of statisticians in academic life, in government work, and in business.'[5]

These twin developments, which were to become major formative influences in our lives, are, of course, intimately connected since—for reasons we shall later examine—the social sciences increasingly tended to work through statistical techniques and to present both evidence and conclusions in numerical forms. This development has been very recent and very rapid.

'Seventy years ago,' points out Professor Richard Stone, there were 'only about 50 books and articles on mathematical economics in the world which were worthy of the name. Now ... not only in economics, but in all the social sciences mathematical books and articles appear by the thousand every year ...'[6] It seems odd now to recall that the statistical breakthrough in economics making 'national—or social—accounting' possible did not occur (in America) until just before the Second World War; that the system of economic aggregates, national income and so on, was worked out in the war years, and that the British Labour Government's postwar annual Economic Surveys, like the American Government's *Economic Reports to the President*, had not so very long ago an air of novelty.

However, with the timely aid of the electronic computer, economic forecasting was quickly added to economic accounting, and by 1961 the US Bureau of the Census was publishing a monthly report which included twenty-eight series of economic indicators, comprising 14 'leading' indicators, 8 'coincident' indicators, and 6 'lagging' indicators plus sixty other economic series predictive of 'turning-points'.

Confusion was multiplied in that in Britain, and some other Western countries, as well as in the United States, these data were not handled by a single government office, but by a range of official and semi-official agencies and rival colleges of augurs, issuing portentous—and often mutually contradictory—numerical verdicts at frequent intervals. Thus it was little more than routine when in London the *Guardian* one day in September 1969 ran a front-page lead story boldly headlined GRIM WARNING ON BRITAIN'S GROWTH RATE—being a conjectural estimate from the National Institute of Economic and Social Research looking for a decline of the Growth Rate to 2·2 per cent in the following year.

Such was news—as news was now made. Nor were such

authoritative statistical 'indicators'—grim or otherwise—confined to economic activities. The meters were now ticking away everywhere from the bedroom to the Cabinet Room. If we had not, in T. S. Eliot's phrase, measured out our lives 'in coffee-spoons', we certainly measured and monitored them with an unprecedented minuteness in percentage points, correlation ratios, predictive scores and running indices of many varieties. The once shadowy force of Public Opinion was now weighed, neatly packaged, and delivered almost daily. Governments which had once felt reasonably secure between elections now saw every fluctuation of their week-to-week fortunes instantly revealed by the seismographs of the professional pollsters. Prime Ministers and Presidents shared with pop stars and TV programmers the roller-coaster of the 'ratings'. All too often the young were switched into their life-tracks in early childhood by the meticulous accountants of the Intelligence.

In advanced industrial countries life was now enclosed by a series of potent and inescapable statistical artefacts—the Gross National Product, the Growth Rate, the Dow Jones or FT Industrial Average, the Crime Rate, the Balance of Payments, the Cost-of-Living Index, the Government's Popularity Lead or Lag, the Industrial Production Index, the index of inflation which the Americans so picturesquely call 'the GNP price deflator'. ... Highlighted month by month, their mystic movements and permutations loomed over the ordinary human lot, much as sacred mountains and spirit-filled trees once dominated the life of primitive man and filled it with vague menace.

From the main statistical artefacts which form, so to speak, the panorama of our life and times, derive many others, ranges of foothills which nevertheless appear to tower high enough in the perspective of those beneath. There is, for instance, that cloud-capped peak called Productivity, permitting a great many complicated triangulation exercises, frequently dubious in the extreme on examination, yet presented for all that with the aplomb of the Recording Angel. Much favoured in the British Press are the so-called 'League' tables—the Growth League, Productivity League, the Share-of-World-Trade League ... in which civilisation takes on the appearance of an endless international statistical relay race.

BRITAIN 14TH IN STANDARD-OF-LIVING LEAGUE, ran a typical

London newspaper headline, implicitly inviting Britons—once again—to hang their heads in shame. The genesis of this particular socio-statistical artefact is perhaps worth a second glance. It is derived by the simple yet well-established process of dividing each country's Gross National Product by the number of its inhabitants, and arranging the resultant figures in descending order. It is an exercise which gains authority from the massive and undeniable weight of the 'GNP' itself, an artefact which lends an air of gravity and omniscience to so many political speeches and editorials.

Unfortunately, the Gross National Product is justly named: it is gross indeed, a vast agglomeration of the like and unlike, made possible only by the fact that monetary values can somehow be clamped on all its items, thus endowing them with the supreme virtue of being available for adding up. To the dedicated economic statistician it represents, of course, a challenge; but to the layman it may appear rather as a most exhaustive device for the multiplication of error—and not only to the layman, if Professor Oskar Morgenstern's devastatingly frank study, *On the Accuracy of Economic Observations*, is to be credited. In an investigation into margins of error in National Income accounting, one of the leaders in the field, Professor Simon Kuznets, broke down the national income (alias GNP) aggregate into no fewer than 520 constituent cells. Then he and his colleagues set out to estimate the probable degree of error in each, taking into account each of its many elements. The final estimated weighted margin of error came to 20 per cent, which was halved to allow for that providential statistical effect, self-compensation by the cancelling out of random 'plus' and 'minus' items. The result was an *average* margin of error of 10 per cent—the equivalent, Professor Morgenstern points out, of three times the value of America's entire export trade.[7]

But whatever the current degree—and mass—of statistical error,* the Gross National Product figure has nothing to say

---

* Some of Professor Morgenstern's colleagues have found him over-severe. But in its 1956 system of 'reliability rating' the British Central Statistical Office grades the GNP figure—'with 90 per cent confidence'—whatever that may mean—at a margin of under 3 per cent plus or minus. The profits, rent, and self-employed income items were given a margin of error of 3–10 per cent, plus or minus. (*National Income Statistics—Sources and Methods*, 1956.) Applied to the calculation of the Growth Rate, such a range of error could still, it seems, work fair havoc. See Chapter 9.

about the manner in which the national income is in fact allocated between rich and poor, between public and private uses, between exploitation and service. And this lack of discrimination is transferred to the concept of 'the standard of living', today held up as the supreme—indeed almost the sole —goal of human endeavour. Whether measured by consumer expenditure or by enumeration of certain goods—refrigerators, television sets, cars—the American standard of life has been held out with great effect as a unique achievement for the emulation of the rest of the world. Yet this form of accountancy neglects much that is surely crucial in any civilised standard of life: the scope and cost of provision for ill-health, the security of the streets by day and night, the manners of the police, the quality of public transport, garbage collection, town-planning. It omits certain other statistics: that, for instance, America comes 13th among the nations for infantile mortality, and 22nd in the life-expectancy table for a boy of ten; that there are forty-eight times as many gun-murders a year as in the roughly equivalent combined populations of West Germany, England and Japan; that almost one in every five young men is rejected for military service on health grounds, or that there are 5,000 small towns without a doctor.[8]

It hardly matters. Buttressed by endless, repeated calculation, immaculate in the halo of statistical objectivity and usage, such 'concentrated' single-number indicators or artefacts are, in effect, as unquestionable as the barometer on the apartment wall. The 'standard of living' figure begs only a few more questions than the others. Consider, for instance, that now global cynosure and supreme index of national virtue, the Growth Rate. Leaving aside, for the moment, the not inconsiderable question of the degree of its accuracy (widely assumed to be total), international comparisons depend enormously on the selection of base-line and period, and blandly ignore such key factors as the availability of large pools of docile immigrant or underemployed labour. Again, the much underlined 'Diminishing Share of World Trade' enjoyed by Britain is still consistent with rising absolute amounts, no way having been found of keeping the world frozen in its nineteenth-century posture without industrial Japan and scores of new nations.

Nevertheless, the 'falling behind' syndrome with all its attendant neuroses is perpetuated.

Of course, statisticians commonly disclaim responsibility for
the indelicate use made of their delicate calculations. No one
is more acutely sensitive to the limitations of their figures than
they. Has not statistics been described as 'the science of error'?

'It is rare to find a piece of statistical information that is
"perfect" in the double sense of being observed without error,
and being precisely suited to its purpose,' report C. F. Carter
and A. D. Roy, the authors of an authoritative investigation
into the accuracy of economic statistics.[9] However, this
fact merely puts the dedicated statistician on his mettle. While
agreeing that 'not even the most subtle and sophisticated
analysis can overcome completely the unreliability of basic
data', Professor R. G. D. Allen nevertheless instructs economist
students that they should not therefore 'discard imperfect data
if nothing else is available ... the best can always be made of a
bad job.'[10] For as Carter and Roy argue in their report, 'fitful
gleams of moonlight are a great deal better to walk by than
complete darkness.'

The only trouble about this is that 'fitful gleams of moonlight' is scarcely an adequate description of the state of our
statistical illumination in the West today; what we are getting
is more like the flashing beacons of a convergent horde of fire-
engines, ambulances and police-cars, each half-blinding us with
its range of numerical indices and indicators, purporting to sum
up, neatly, scientifically and conclusively, our economic,
moral and psychological past, present and future.

It is not of course here being suggested that our complex
modern society does not need the powerful analytical and diagnostic aid of statistical science. It is not suggested that America
does not face the serious recurrent problem of the swings of the
business cycle; what is suggested is that this obstinate phenomenon—in which the vagaries of mass psychology manifestly
play a large, if elusive, part—is not necessarily best dealt with
by focusing national attention hypnotically on, say, decimal
point percentage quarterly changes in the Gross National Product.

It is not suggested, either, that Britain has not had a balance
of payments problem, or that the public should have been kept

in ignorance of it. What is doubted, particularly since these crises are not wholly due to British moral turpitude but may be triggered by wholly extraneous circumstances,* is whether, in an inevitably volatile situation, external and internal confidence is best served by concentrating the headlines on monthly fluctuations of the visible 'Trade Gap', even, as the statisticians say, 'seasonally adjusted'.

Particularly in view of the frequency with which not merely the monthly, but also the quarterly and annual, figures are subsequently officially 'revised', on occasion even to the extent of turning what had been a deficit into a surplus. It was indeed a remarkable comment on the inroads which statistical neurosis had already made that it was perfectly possible in Britain—and no doubt not only in Britain—for at least a minor financial crisis to be triggered off by obsessive publicity focused on clear-cut and menacing deficits subsequently found to have been statistical figments.[11]† Certainly, there was more than mere whimsicality in the remarks of the British Chancellor of the Exchequer who was given to explaining to foreign audiences that there was no balance of payments problem in the nineteenth century because there were no balance of payments statistics.‡

In his Presidential speech to the Royal Statistical Society of London, Sir Paul Chambers, a former Chairman of Imperial Chemical Industries, and himself a statistician, spoke of his 'firm belief that much self-criticism of our [Britain's] postwar

* As for instance by the disturbances caused by the rumours that Germany was about to revalue the Deutschmark in 1957.

† The aggregate balance of payments deficit 1946-8, originally proclaimed to the world as more than £1,100m, was later revised to £737m; the 1960 deficit of £334m reduced to £258m and so on. Revisions may also be revised. The fallibility of the very complex processes which end up in the neat public totals was demonstrated in mid-1969 when the Board of Trade found that, owing to a technical oversight, it had been under-recording exports to the average extent of £10m a month for the preceding five years; for the previous year the under-recording was no less than £130m. All of which bears out the contention of some economists (e.g. F. V. Meyer, in *The European Review*, 1967; Norman Macrae, *Sunshades in October*) that British balance of payments figures were generally too pessimistic.

‡ According to Norman Macrae (above): 'Nobody ever attempted to work out this statistic before the 1930s; the old-time policy makers only dimly realised that it might theoretically exist.' They looked to the movement of the Gold Reserves instead—and to that extent had at least one fewer in the roll of agitations which their descendents multiply with such dedication.

performance is based on misinterpretation of statistics'. He added: 'The practice of picking out particular series—short or long—and coming to a judgment that performance has been good or bad without reference to the whole picture of economic policy and performance is widespread and pernicious.'[12]

Indeed, it extends far beyond the bounds of economics, and is a critical part of the climate—of the hypochondria—of our times.

The Disease is the Diagnosis

The psychology of individuals has had much study, but the psychological processes of whole societies have received little attention. Their economic behaviour, for instance, is assumed to be uniformly rational and independent of passing mood. Yet anyone who has watched a competitive team game is aware of the importance of elusive psychological factors, of the critical role of confidence and of the team's immediate idea of itself. He will also have noted how often this can be changed dramatically by changes in the score. Sometimes, indeed, the score is the man.

Proliferating in these years, manning—most notably in the United States—a vast research industry, the social sciences have raised social self-consciousness to wholly unprecedented levels. What is perhaps more important, they have given it both continuity and a daunting 'objectivity' by presenting their findings as numerical indices—an endless and ever-widening succession of 'scores'. Increasingly, important articles in today's newspapers—on political and social as well as economic topics—tend to consist of rationalisations of these statistical indicators, explanatory comment on the current scores.

While daily—indeed almost hourly—the social sciences thus present us with giant diagrams of ourselves as dissected and calibrated by some sector of social science, what they do not give us is a similar scrutiny of the social sciences themselves and the impact of their slide-rules on society—upon us. There are solid prudential reasons for this reticence on their part, as will emerge. All the same, the omission is unfortunate. There can be few more important subjects today.

At this stage social scientists might do worse than take a leaf from the book of those *other* doctors—the doctors of medicine. Standing somewhere between the social and physical sciences, modern medicine has become increasingly aware of the psychological factors in disease. Not only does it recognise a wide range of psychosomatic illness, but it has also acknowledged the category of Iatrogenic Disease—ill-health or neuroses brought on by the doctor himself.

The cause may be simply an error in diagnosis caused by concentrating on one symptom and thus missing another, perhaps less readily measurable. Or it may be a misjudgment in treatment, arising perhaps from the side-effects of some potent drug. More characteristically, Iatrogenic Disease is defined as a disease 'unwittingly induced by a physician in a patient by autosuggestion based on the physician's words or actions'. These may amount to no more than some unguarded inflection of the voice or some fugitive fragment of medical jargon. Diagnostic measuring instruments have proved a particularly fertile source of such disquiet. Blood-pressure readings can notoriously contribute to raising a susceptible patient's pressure still further, and we are told that 'some teachers advise against telling the patient what is learned from the sphygmomanometer and even against using it too frequently'.

It has indeed been rather cynically said that each new medical monitoring instrument produces its crop of new diseases, some at least of which suffering humanity has managed to do well enough without hitherto. Certainly this is a thought from the MD's office that the PHDs among the IBM machines might pause to ponder.

All Done by Numbers

This book will suggest that the progress towards social hypochondria, iatrogenically induced, is already dangerously advanced. There is much that points that way—the manic swings from euphoria to despair triggered by the stock market averages (and certified by the 'Confidence Index'), the narcosis of the endless opinion polls, the nervous tic of the economic indicators, the anxiety neurosis generated by intelligence and personality rating, the withdrawal from reality found in chartists and other trend-watchers, the vivid delusions born of cost-benefit arithmetic or 'standing room only' projections of 'exponential' population growth, the *anorexia nervosa* brought about, particularly in the young, by the rapid expansion of the new social 'science' of psephology. (Who would have thought that the magnificent simplicity of one man, one vote, would come to this?)

Today's doctors of medicine are accustomed to spend almost as much time weighing the 'side-effects' of some potent 'wonder drug, as they are in considering its direct therapeutic qualities. They have learned from sometimes bitter experience. Today refined statistical and mathematical techniques have brought to some of the social sciences a breakthrough reminiscent of that in medical science. But the social scientists still appear stuck at the 'wonder-drug' stage.

Study of 'side-effects'—some of them certainly potent—is clearly overdue.* But to muster the necessary detachment cannot be easy in complex industrial societies such as ours, societies inevitably as dependent on statistics as a blind man on his stick, societies in which 'the institutions' may now mean not Parliament or Congress or the Supreme Court but the central core of banks, great insurance companies, investment trusts, finance houses, which live and have their being by statistical indices and indicators. Figures are faceless and incestuous. In a society dominated by them, the accountancy of intelligence tests and the psychometricians meshes as readily with that of insurance actuaries and investment managers as with the ratings and class ratios of the mass media or the cold calculations of psephologists and other practitioners of 'scientific politics'. And to isolate and dissect any particular effect can be a difficult exercise when the whole system of complex statistical interaction, here called the Numbers Game, has come to resemble some vision of perpetual motion with no evident directing hand, no visible programmer. It is confidently asserted that its purpose is to afford citizens the greatest freedom of choice. But it is

---

* One of the few areas in which they have been seriously considered is that of intelligence testing (Chapter 11). See also *Artifact in Behavioural Research*, edited by Robert Rosenthal and Ralph L. Rosnow, New York and London, 1969.

hard to tell the driven wheels from the driving. All we seem to know is that somewhere in the middle of this great train of gears stands man, the consumer, the shareholder, the wage-earner, nervously watching the numbers as they come up ...

It is the nervousness that concerns us here. The diagnosis of so complex and persistent a neurosis must examine not only its onset and course, but also the inner mechanism of the causative statistical 'side-effects'—from the false confidence born of phoney numerical precision to the disturbances of vision brought about by an excess of mathematical abstraction. In doing so, it will be obliged to consider the conditions in the worlds of the mass media, the social sciences and merchandising which greatly intensify these side-effects, forming, so to speak the cultures in which they proliferate.

But first it will be necessary to consider the origins of some of the statistical techniques which are now so much a part of the fabric of our society. The history books have little or nothing to say about their arrival on the scene, yet they have been among the most potent inventions of our time, as socially dynamic as the steam-engine or the aeroplane.

FROM MALTHUS TO KINSEY:
THE MATRIX OF NUMBER

Bless us, divine number, who generatest gods and men, O holy *tetraktys* that containeth the root and source of the eternally flowing creation.
                    Invocation of the pupils of Pythagoras
                    to the number Four, c. 520 BC.

... We arrive at a figure of 3.27 per week for the total sexual outlet of the average white American male under thirty years of age (Table 40). For all white males up to 85 the corrected mean is 2.34 per week. The latter figure is lower because of the inactivity of the older males.
                    Kinsey Report on *Sexual Behaviour in the*
                    Human Male, 1948.

Somewhat earlier than the Council of Economic Advisers or the National Institute of Social and Economic Research, Pythagoras asserted that number ruled the universe. However, he demonstrated the fact somewhat more directly than later statisticians, simply endowing numerals with moral qualities —one for reason, two for opinion, four for justice, five for marriage—the last being formed from the union of the first male number, three, with the first even, or female, number two. Some eight hundred years later St Augustine explained that if 'God created all things in six days it was because this number is perfect'—its divisors, one, two, and three, totalled itself. The accident that both in Greek and Hebrew letters were also numbers inspired the esoteric numerical exercises of the Cabala. The triumph of Achilles over Hector, for instance, was explained by the fact that the letters of Achilles' name totted up to 1,276 whereas all Hector could muster was 1,225, a demonstration that was at once elegant, conclusive and anti-cipatory of much that was to come in later centuries.
    For whatever their form, numbers were to retain their spell and curious quality of exacting acceptance—or suspension of

disbelief—into the scientific age, as Mr Heinz doubtless grasped
when he plastered wallsides and cliffsides all over America with
the legend '57 Varieties', no less than Senator Joe McCarthy
when he reported 205—or 57, or 81, or 10, or 116, or 121—
card-carrying Communist Party members in the US State Department. If the number varied down the months, it was always
crisp and definite, and far from occasioning disbelief, soon 'had
half the nation disputing over precisely how many Communists
there *were* in the State Department'.[1] Oddly enough, exactly
the same 'curious precision' is to be found in the accusation
of the witches at Salem in 1692: there were 307 witches and
105 'young wizards'.[2]

The progression from the cabalistic to the scientific in
statistics—or from the older magic to the new—has been slow
and long and resisted much of the way. Thus when in March
1753 a Bill was introduced into the British House of Commons
proposing the first national census since William the Con-
queror's Domesday Book, letters of protest and foreboding
poured into Westminster from all over the country. It was recalled that when, tempted by Satan, King David had numbered
the people, the Lord had punished him by sending a pestilence.
One member of Parliament described the Bill in the House
debate as 'totally subversive of the last remnants of English
liberty ... And what purpose will it answer to know where the
kingdom is crowded and where it is thin except we are to be
driven from place to place as graziers do their cattle? If this is
intended, let them brand us at once; while they treat us like
oxen and sheep, let them not insult us in the names of men.'
Declaring that he could 'see no use in the Bill' another Member
found it 'merely a satisfaction for those gentlemen who love
to deal in political arithmetic'.[3]*

He need not have worried; the House of Lords threw out the
'slavish' Bill, and it was almost another half century—and ten
years after the first national census in the United States—before
the true-born Englishman was subjected to the indignity of
being 'numbered'.

* The forerunner of economics, political arithmetic, was described by the
seventeenth-century political arithmetician, Sir Charles Davenant, as 'the
art of reasoning by figures upon things relating to government'. Davenant
himself produced a book with a still highly topical title: *Essay on the
Probable Method of making a People Gainers in the Balance of Trade* (1699).

For all that, it was the 'political arithmeticians' who were to have the last laugh. By 1790 Edmund Burke, with his ear to the ground, was already lamenting the onset of the age of 'sophisters, economists and calculators'.

'The Stedfast Order of the Universe'

If a development which so powerfully shaped our times, and today in so many ways rules our lives, has been largely neglected by historians, it is because the Statistical Age did not so much dawn upon the world as steal up on it. The science of statistics was more than two centuries in the making. Its lineage is curious and complex. Demographers and 'political arithmeticians', astronomers seeking the causes of their observational errors, pure—and impure—mathematicians, actuaries refining their life tables, doctors of medicine and public health, geneticists, ethnographers and psychologists, all made their contributions, building the armoury of techniques. But the main early sources of this reputedly most neutral and colourless of sciences sprang from two very human centres of powerful emotion—Death and High Stakes, more immediately, the plague pit and the gaming table.

The first stream, flowing from Mortality, originated in England; the second, deriving from Chance and Probability, came from Italy and France.

In 1532 the Lord Mayor of London was asked by the Privy Council to furnish it with particulars of the numbers of citizens dying of the Plague. After the Great Plague of 1603 these returns—the Bills of Mortality—were made at weekly intervals, and included christenings as well as burials. From 1629 onwards the cause of death also was listed, providing enough data to enable John Graunt, a City haberdasher and former captain in the City militia, to bring out, in 1662, a little book entitled *Natural and Political Observations upon Bills of Mortality*.

The raw materials of this pioneer demographer were raw indeed. 'Ancient matrons' in each City parish were required to 'examine the corpse and report the Distemper' which, in a vivid list, might include 'Griping in the Guts', 'Grief', or, simply, 'Frighted'. Nevertheless, assiduously working over the Bills, Graunt was, as he puts it, able to come upon 'some truths and not commonly believed opinions'. He found, for instance, that Nature maintained a rough balance between the sexes, with male births slightly exceeding female—a novel idea at a time when population was estimated from the number of chimneys and it was commonly believed that there were three women to every man. Further social regularities were laid bare, and out of them Graunt constructed the first crude Expectation of Life Table which showed that 36 per cent of those born alive in London would die before reaching the age of six.

Highly faulty as it was, John Graunt's table, linking the rate at which people die to the numbers of survivors at various ages, was a major social invention which, after refinement by many hands, was finally to provide the foundation of the vast life insurance industry. In 1690, for instance, Sir Edmund Halley, the Astronomer Royal—a brilliant mathematician who had identified Halley's Comet by means of a calculated average—turned his hand to the construction of an improved Life Table built on the recorded vital statistics of the Saxon city of Breslau which were both fuller and more normal than London's.

Meanwhile in Europe it was the fall of the dice and the chances of the card-table rather than the odds of mortality that was slowly sharpening the calculus of Probability. Much gaming expertise spoke in the mathematician Cardano's book on *Games of Chance* with its crucial chapter 'On the Cast of One Die', and the great Galileo himself responded to an appeal from local gamblers with a short treatise, written in 1620, analysing the probabilities when three dice were cast together. In France, Pascal and Fermat in turn took up the problems of the dice; in the Netherlands Huygens produced in 1657 his *Calculations in Games of Chance*; and in Basle the Professor of Mathematics, James Bernouilli, rounded out the theory with his *Art of Conjecture*, including a section on permutations and combinations.

A young Huguenot mathematician, fleeing from France after the revocation of the Edict of Nantes, carried these preoccupations across the Channel. Unable to get a University post in England, Abraham de Moivre had to resort to making a living advising both gamesters and speculators in lives and annuities, and in 1718, he incorporated some of the results of his experience in a celebrated work, *The Doctrine of Chances*. Dedicated to his friend, Sir Isaac Newton, de Moivre's book boldly proclaims the statistician's fundamental article of faith that Chance is not *merely* chance, but that 'there are in the constitution of things certain laws according to which events happen to preserve the stedfast order of the universe.'

In the mathematically refined life tables of de Moivre and his successors were celebrated not only 'the stedfast order of the universe' but the fertile union of the mathematics of Probability and observed social data on which not only the potent social science of statistics but the great cash pyramids of the life insurance industry were to be built. In 1762 the first life insurance company to operate on a scientific basis, the Equitable, was founded in London.*

But it was not until the last decade of the eighteenth century that a strange new word, 'statistics,' arrived in England from Germany. It was imported by Sir John Sinclair, the first President of the Board of Agriculture, an assiduous Scot who anticipated the first English census by a decade, gathering materials for a comprehensive survey of Scotland from the parish ministers in every county and attracting the favourable attention of George Washington across the Atlantic. The first of the survey's 21 octavo volumes appeared in 1791 under the title *A Statistical Account of Scotland*. The German term statistics, explained Sinclair in the Preface, did not quite match his own purpose, since statistics, German-style, were concerned with the materials of the political strength of a state, whereas he himself was more interested in 'the quantum of happiness enjoyed by the inhabitants and the means of its improvement; yet as I thought a new word might attract more public attention, I resolved to adopt it.'

This expectation was to prove correct: the fascination of 'statistics' turned out to be perennial.

The World of the Reverend Thomas Malthus

If the development of statistical techniques raised some major material monuments, its effects on social, political and

---

* The first American commercial life insurance company to use American mortality tables was the Pennsylvania Company launched in 1812. But the far-reaching union of life insurance and American salesmanship did not come until the mid-fifties.

moral ideas and institutions were hardly less far-reaching. Gathered from the past, statistics increasingly formed the mould of the future. The men of figures were quiet, almost invisible, men; they infiltrated rather than conquered, yet with each successive decade their dominion widened, their empire grew. In this chapter we will look at three of these men who, over the last century and a half or so, have by their use of statistics changed history. They are the Reverend Thomas R. Malthus who launched his statistical proposition in 1798; Charles Booth, whose main work began in 1889; and Dr Alfred Kinsey, whose name rang around the world in 1948. Between whiles we will refer to two major pioneers of statistical technique whose work lies behind the statistical artefacts that guide our lives; a Belgian, Adolphe Quetelet, and an Englishman, Francis Galton.

The statistical proposition of each of the three men—a mathematically minded young Surrey curate, a rich Liverpool shipowner, and an Indiana zoologist, all in their very different ways illustrate the immense social potency of supposedly inert statistics. Each marks a new forward movement in the expanding empire of Number. But even today it is probably the earliest of the examples which most vividly brings home the blinding, mind-paralysing power a simple statistical formula—correct or incorrect—can possess.

That Malthus's famous 'law' of population has so long continued to grip men's minds owes much to its aura of relentless objectivity. It is particularly fascinating therefore to find that its origins at least were highly subjective: it was a son's overdue declaration of independence from a strong-minded father. Malthus *père*, a philosophical country gentleman, was an admirer of the early French Revolution and a disciple of Condorcet and Goldwin, sharing their faith in the perfectibility of man. To his parent's tireless espousal of the large possibilities of Nurture when shaped by Reason, the son riposted, in classic fashion, with the divine tyranny of Nature. And since at Cambridge he had been ninth Wrangler he placed at the centre of his counter-revolutionary polemic a mathematical idiom: population advancing by geometric progression, food supply merely by arithmetic progression. Utopia and parent were put down at one blow.

But while the Reverend Malthus sternly rebuked his father

and his likes for 'believing what he wished without evidence', his own theory too was, in fact, almost wholly *a priori*, a brilliant hunch. In his slender *Essay on the Principle of Population* of 1798, the only real evidence he adduces for his statement about the propensity of population to double every 25 years derives from some figures for four provinces of New England, and even here he simply assumes that immigration and emigration cancel out without taking in account the differences in fertility at different ages. The other blade of his proposition, the arithmetical advance of food supply, is little more than a 'common sense' hypothesis.

Malthus himself said that he wrote 'on the impulse of the occasion, and from the few materials which were then within reach of a country situation'. Only after the furore which followed publication did he embark on travels in Europe in search of substantiating evidence. This he continued to set out in self-justifying new editions for most of the rest of his life; the 55,000 words of the First Essay became the three volumes and 250,000 words of the fifth edition.

Nor were Malthus's ideas and material original. As he himself pointed out, he owed much to Adam Smith, to David Ricardo, and to two other philosophical clerics, the political arithmetician, Dr Richard Price—constructor of yet another famous life table—and the German pastor, Johann Peter Süssmilch. Süssmilch was the archetypal German statistician. For him, the Almighty was the 'infinite and exact Arithmaticus ... who has for all things in their temporal state their score, weight, and proportion'. His book, *Die Göttliche Ordnung,** was the first systematic treatise on population, and the 1,201 pages and 207 tables of the 1765 edition provided a wonderful quarry for his fellow curate of Surrey.

But the originality of Malthus did not lie in his facts. When Coleridge, who read one of Malthus's later, swollen, editions scribbled in the margin: 'Verbiage and useless repetition', he had a point. For what matters in Malthus, what makes his name still vividly remembered when those of such industrious statisticians as Süssmilch and Sinclair are long forgotten, is not his

* Translated, the full title reads: *Reflections on the Divine Order in the Mutation of the Human Race as indicated by its Birth, Death and Propagation.*

apparatus of evidence, or even his philosophy, but the searing clarity of his central, mathematical proposition.

Of his postulated parallel geometric (population) and arithmetic (food supply) progressions Malthus tells us blandly, 'A slight acquaintance with numbers will show the immensity of the first power in comparison with the second.' But he is not content to leave the matter there. He parades the figures before us:

> ... the human species would increase as the numbers 1, 2, 4, 8, 16, 32, 64, 128, 256; and subsistence as 1, 2, 3, 4, 5, 6, 7, 8, 9. In two centuries the population would be to the means of subsistence as 256 to 9; in three centuries, as 4,096 to 13, and in two thousand years, the difference would be almost incalculable.

Where is the romantic reformer who can stand against that? The precision of these doom-laden figures might be a wholly false precision. The basic proposition might be wrong—as indeed events were finally to show. It did not matter. The relentless statistics of Malthus were to dominate the ruling-class vision in Britain for three-quarters of a century, causing kindly men to support a social system of great inhumanity and sanctifying the greed and self-interest of those less kindly.

Malthus's 'law' brought a statistical formula or a mathematical idea to bear on a great mass of data. But the mathematical idea was a good deal more lucid and therefore compulsive than the disorderly data which were supposed to have formed it. The proposition that emerged combined both the spell of number and the authority of 'science'. It was—and is —a potent amalgam.

Then, again in a way that was also to become increasingly familiar, Malthus boldly undertook a massive extrapolation, continuing his geometric curve hypnotically upwards and onwards into the blue—or, in this case, murky—yonder. The aura of determinism—divine, of course, in the case of Malthus —extends to the outlining of a system, an early socio-economic 'model' complete with constants, variables and a built-in automated equilibrium, the latter borrowed from Adam Smith.

Thus, a constant was 'the passion between the sexes' ... 'in every age ... so nearly the same that it may always be considered in *algebraic language* as a given quantity.' Other constants were the area of the soil, and the need of the human body for a certain minimum sustenance. Variables were the rate of population growth, disease, migration, war, famine, death ...

Never have both the potency and limitations of statistical models been more sharply exposed. Malthus produced his system with his mind rooted in the soil and in the Law of Diminishing Returns that then loomed so large. But even as he wrote industrialisation was gathering pace, heralding a new world. Some of Malthus's constants shortly turned out to be highly variable. Liebig demonstrated the immense stimulus which chemical fertilisers would exert upon plant growth; Gregor Mendel in his monastery near Vienna inaugurated the science of plant genetics; in America the steel plough and combine harvester foreshadowed the mechanisation of agriculture. And if 'the passion between the sexes' remained undiminished, its demographic consequences were increasingly averted by birth-control, practised not from fear of starvation, but in order to protect a 'standard of living'.

Malthus had given the world another 25 years before the grim food squeeze set in. In fact, more than a century and a half later, in the Western World at least, food supply has more than kept pace with a leaping population.

Frank W. Notestein, President of the Population Council in America, sums up: 'What to Malthus were the possibilities of minor displacements from equilibrium became in the modern West more than a century of progressive release from the fundamental constraints with which he was preoccupied.'[4]

Yet even in a Britain become the 'workshop of the world' the vision of Malthus remained compulsive. He himself seems to have been a particularly nice and kindly man, and to himself and to others his statistical system appeared as an observation of total scientific objectivity. In fact, as models will, it constructed and consolidated a world as well as reflected one. Its masking of political self-interest was all the more effective for being so sublimely unconscious; the political philosophy it embodied appeared inevitable, God-given, caused, rather than causing.

This too will be seen again.

Such systems tend to be self-sustaining. Charles Darwin himself wrote that it was Malthus's sharp statistical vision of the human swarm's struggle for sustenance that suggested to him the basic mechanism behind Evolution, the driving force of Natural Selection—'the Survival of the Fittest'. And this, in turn, bred 'Social Darwinism', *laissez-faire* sanctified by the clear need for the weakest to go to the wall, which, improbably, reincarnated the vision of Malthus in the America of Andrew Carnegie and John Pierpont Morgan.*

Equally, the nightmare arithmetic of Malthus inspired the schemes of Edward Gibbon Wakefield for the colonisation by organised emigration of Australia and New Zealand and provided material for the system of Karl Marx.

Such is the power of 'divine number'.

Charles Booth Draws 'the Poverty Line'

In 1834 Thomas Malthus, then long-time Professor of History and Political Economy at the East India Company College at Haileybury, became a founder member of the Statistical Society of London. He died in the same year, but a fellow founder, Charles Babbage, was already foreshadowing a statistical era of a more dynamic order than any the theologically minded Malthus could have dreamed of. Babbage was the inventor of two vast mechanical 'calculating engines'—a 'difference engine' and an 'analytical engine'—and was to spend most of his life and fortune on the effort to perfect them. He described the 'analytical engine' as 'the Engine eating its own tail'—a vivid anticipation of the principle of 'feed-back' which was to play a critical role in the evolution of the Numbers Game—and his 'calculating engines' in fact embodied the basic ideas and many features of the electronic computer developed after the Second World War.

However, Babbage's invention failed for lack of a suitable driving power, and 'statistics' continued to be seen largely in terms of the collection and tabulation of masses of numerical facts. In London the new Statistical Society adopted as its crest a wheatsheaf with the motto, *aliis exterendum*—'to be threshed

---

* 'For cases of extreme distress', Malthus himself did, however, sanction the admission of 'the weakest' to the 'county workhouse' while stipulating that 'the fare should be hard, and those who are able obliged to work'.

out by others'. 'The most essential rule,' pronounced its Report for 1836/37 (which surely must have provided Dickens' inspiration for Mr Gradgrind, the dedicated statistician of *Hard Times*), 'is to exclude carefully all opinions—to confine attention rigorously to facts, and facts as far as may be possible which may be stated numerically and arranged in tables.'

There was little idea yet of using statistics purposefully to construct tools of research, although there was one brilliant exception in the work of Dr Pierre Louis, a Paris doctor, who in 1835 exposed the uselessness of the orthodox treatment of blood-letting in inflammatory conditions like pneumonia by setting his hospital case results against those of a non-bled control group.

Dr Louis's medical colleagues derided his 'Numerical Method'* which they seem to have felt an affront to their professional dignity. But the bacteria of the crowded and fast-growing towns were no respecters of professional dignity, and the power of the statistical microscope was slowly established. In London in 1854 when such infections were still generally believed to be due to 'poisonous miasmas', Dr John Snow, a local general practitioner, halted a deadly outbreak of cholera in Soho by the simple expedient of charting on a street map the houses where the disease had struck. His map showed that all the 500 deaths which had occurred in ten days had been in houses within 250 yards of one water-pump. The pump handle was removed, and the outbreak ceased.†

It was to be another thirty years before the water-borne bacillus of cholera was discovered by Koch. But already the

---

* A remarkable anticipation of 'the Numerical Method' occurred in Boston, Massachusetts, in 1721, when the Reverend Cotton Mather and Dr Bolyston, surveying smallpox cases, reported to the Royal Society of London that 1 in 6 of all who took the disease died, but only 1 in 60 of those inoculated by Jenner's method.

† Cf. *The Lancet* in 1853: 'What is cholera? Is it a fungus, an insect, a miasma, an electrical disturbance, a deficiency of ozone, a morbid off-scouring of the intestinal canal? We know nothing ...' In 1856 Dr Snow's diagnosis was confirmed by a new statistical foray into South London, where the pipes of two rival water companies served the same streets. The Southwark and Vauxhall Company took its supply straight from the Thames at Battersea; the Lambeth Company had three years earlier switched to a point up river above Teddington Lock. The Southwark Company's clients had a cholera death rate of 130 per 10,000; the Lambeth Company's 37 per 10,000, although its cholera rate had formerly been as high as the Southwark's.

new public health doctors were demonstrating, statistically, that dirt, poverty and disease were companions. In 1839 a young Shropshire doctor named William Farr, who had encountered 'the Numerical Method' of Dr Pierre Louis when a student in Paris, joined the recently formed General Registry Office of England. And from then until 1880 when Dr Farr retired, the annual reports of the Registrar-General directed a probing statistical searchlight into the darker corners of the booming towns of industrial Britain, revealing infant mortality, malnutrition and epidemic disease. Farr did much to make vital statistics internationally comparable—a sharp new index of human welfare.

Unbidden, what Thomas Carlyle called 'The Condition of the People Question' was belatedly rising to the surface of the national consciousness. In 1883, a report of the London Congregational Union, *The Bitter Cry of Outcast London*, sold out edition after edition. Two years later the Social Democratic Federation, led by the wealthy socialist, H. M. Hyndman, organised a survey of working-class London which showed that 25 per cent of its people were living in poverty. And out of California came the American journalist, Henry George, with his bestselling *Progress and Poverty*, and the Message—delivered in person to eager London audiences—that the merciless world outlined by Malthus had, in fact, been constructed neither by the Almighty nor by Nature, but was the work of men—who could change it.

But fear of the multiplying, unregenerate hordes conjured up by the Reverend Malthus' 'progressions' still haunted the cosy middle-class firesides. The Poor Law of 1834, sternly excluding all normal human charity, remained intact. The spell exercised by the statistics of Malthus could be broken only by another, and equally potent, set of statistics.

This was what Charles Booth provided.

But the statistics of Booth, while no less decisive, were of a different order from the statistics of Malthus. Booth was neither a mathematician nor a philosopher; he was a hardheaded, down-to-earth Liverpool businessman. He both harvested his 'numerical facts' *and* 'threshed' them. Certainly he did not announce the totals of his crop before he had even weighed it. His great statistical survey of the *Life and Labour of the People of London* continued from 1887 to 1903, and finally filled seventeen volumes. It forms a towering landmark, both in the history of social statistics and in the social history of statistics.

One thing, however, the *deductive* Malthus and the determinedly *inductive* Booth did have in common. Both launched on the statistical investigations which were to become their life's work for very personal reasons: if Malthus was reacting against his Utopian parent, Booth wished to prove ill-founded what he regarded as the cheaply sensational poverty reports of the Hyndman Socialists, then being splashed in the *Pall Mall Gazette*. For, like most of his class, he fully subscribed to the Malthus-inspired belief that 'the rich are helpless to relieve want without stimulating its sources'.

And yet ... he was vaguely uneasy. He was becoming increasingly aware that, as he wrote, 'the *a priori* reasoning of political economy fails for want of reality. At its base are a series of assumptions very imperfectly connected with the observed facts.' Booth was a Unitarian in religion; his was a notable case of the Nonconformist Conscience in action. He resolved to take a long, hard look—a *scientific* look—at the 'observed facts' of life as they were to be found in the capital.

There was no precedent for such a detailed, statistical survey of the lives of four million people. Malthus, in his first *Essay*, had declared himself hampered because 'the histories of mankind which we possess are histories only of the higher classes'. The masses of great cities such as London were totally anonymous—unplumbed depths on the edges of which the middle classes walked with careful step and averted eyes. 'East London,' wrote Booth 'lay hidden from view behind a curtain on which were painted terrible pictures ...'

He resolved, at the outset 'to make use of no fact to which I cannot give a quantitative value'. For his pilot survey in East London, he hit upon an excellent source. The School Board inspectors, charged with getting the children to school, had an intimate knowledge of their districts, street by street. It was upon their notebooks that Booth built.

In a letter to his assistant, Beatrice Potter, his wife's cousin, he wrote: 'What I want is to see a large statistical framework ... built to receive accumulations of fact out of which at last is evolved the theory and the law and the basis of more in-telligent action.'

He finally devised a framework of society composed of eight socio-economic classes. They ranged from a 'lower depths' of 'casual labourers, street sellers, loafers, criminals and semi-criminals' which Booth labelled 'Class A', through a Class B of the 'very poor, ill-nourished, poorly clad' to Class F, 'the non-commissioned officers of the industrial army', and, at the sum-mit, Class H, 'the servant-keeping class'. Into this frame the families and individuals of each street were fitted, checked and re-checked against household budgets, occupations and earn-ings. At one end there was an 'overcrowding index' of persons per room; at the other, a measure of affluence: servants per family.

Booth's next preoccupation was to locate what he called 'the Poverty Line', a phrase possibly suggested by his familiarity as a shipowner with the Plimsoll load-line. It was a statistical artefact of the greatest importance. Like Malthus's 'progres-sions' it had the sort of mathematical sharpness that can make history.

Booth set the Poverty Line at '18 shillings to 21 shillings per week for a moderate family'—which had the effect of bringing the lower four classes within his definition of poverty. For Classes C and D, he wrote, life was 'an unending struggle', but if they were frugal, and their luck held, they might just make ends meet. Classes A and B, however, were 'at all times more or less in want'.

In thus defining 'the steps of poverty' Booth did not rely wholly on statistics; he spoke with a different sort of authority —the authority of one who has 'been there'. He haunted the streets his statistics dissected. Several times he lodged for a time with a worker's family, sharing their table. His notebooks filled up with vivid details which even now can make one feel the people's poverty. Of Class B families, he notes: 'They go to the shop as an ordinary housewife goes to her canisters; twice a day they buy tea, or three times if they make it so often ... the "pinch of tea" costs ¾d ...'

'The statistical method,' Booth wrote to Beatrice Potter, 'was needed to give bearings ... but personal observation to give life to statistics ... the figures or facts may be correct enough in

themselves, but they may mislead from want of these proper-
ties or from lack of colour.'

It was a potent combination of the sensitively qualitative
and the scrupulously quantitative, and it exploded two *idées
fixes* which had remained rooted since the time of Malthus. The
first was that poverty was an inevitable natural phenomenon,
arising in some vague way from Original Sin or the innate
delinquency of the Lower Orders. Far from being due to 'Drink',
Booth demonstrated that the plight of over half Classes A and
B and over two-thirds of Classes C and D was attributable to low
wages, lack of regular work, or low profits. Nearly all the re-
mainder arose from illness, large families, or old age.

The second Malthusian *idée fixe* was that any improvement
in the lot of the poor was likely to be self-defeating, since they
would at once fecklessly multiply—breeding, like animals, to
the limits of subsistence. Booth's field work established beyond
doubt that, on the contrary, a modest competence was the
beginning of restraint. 'Every increment in income, and especially every rise in the regularity and security of the income in
working-class families,' noted Beatrice Potter, Booth's early
collaborator, 'was found to be accompanied, according to the
statistics, by more successful control of the birth-rate.'

When Booth completed his surveys of the East End and
found there 'one-third of the population sinking into want', he
expected that as he moved on to other parts of London, the
proportion on or beneath 'the Poverty Line' would drop. It did
not. On the giant map of the capital he had prepared, with
streets marked in eight colours, the blue-black streets—denot-
ing the cancerous tissue of poverty—were to be seen creeping
out all over the city. Even in such middle-class areas as Maida
Vale and Notting Hill, the poor streets—almost half—infiltrated behind the stately porticos and terraces. All over the
capital of the world's greatest empire 30·7 per cent of the
people were living in penury.

Booth's report made headlines everywhere. *The Times* called
it 'the grimmest book of our generation'. Booth had set out to
show that the socialists were sensationalists; instead, he had
proved them right. Furthermore, his figures had shown, as
nothing else could, that if guilt there was for this massive
poverty and human deprivation, it could not lie, in the main,

at the door of the poor themselves. A disturbing question was thus silently posed: who, then, *was* to blame?

Booth himself was a Conservative, a lifelong apostle of free enterprise and individual responsibility. Of the Salvation Army's efforts he notes in his Survey: 'The ultimate result of providing food and shelter at uncommercial prices can hardly be other than evil.' But now, shaken by the appalling facts he had brought to light, he modified his position. He worked out, and strenuously campaigned for, a programme of old age pensions of five shillings at sixty-five. When this was denounced by the influential Charity Organisation Society as 'the most outrageous and absurd scheme yet promulgated', a reckless squandering of the national substance, Booth answered his many detractors by claiming that a measure of 'limited socialism' was indeed necessary to make individualism work.

But further than old age pensions, more education, and compulsory work camps for 'Class A', he would not go. The shadow of the Workhouse which had fallen so heavily across English working-class life since the new Poor Law of 1834 was still deemed a necessary stimulant to self-reliance. Yet if the ideas of Malthus lingered, their grip was now broken. The meticulous statistics of Booth possessed a cutting edge that had been absent from the rhetoric of a Ruskin or Carlyle, from the impassioned editorials of a W. T. Stead, or even from the novels of a Dickens or Disraeli. Disraeli had merely written of 'the Other Nation'; Booth had delineated it.

In so conservative a land as England it would be many years before the point was fully taken. Yet in some quarters it was taken almost immediately. Beatrice Potter, the rich young socialite who was Booth's collaborator, drew from the statistics the conclusion that was impossible for Booth himself. She decided that tinkering was not enough. She withdrew to marry Sidney Webb, thereby embracing socialism and the Fabian Society; and in 1909 she signed the famous Minority Report of the Royal Commission on the Poor Law which proposed to attack the causes of poverty, rather than merely to accept and relieve it. Another of Booth's collaborators went on to organise the first state labour exchanges and, later, unemployment insurance; yet another to initiate the Trade Board Acts setting legal minimum wages in sweated trades.

The majority report of the 1909 Poor Law Commission had still sternly ruled that 'the unemployed man must stand by his accidents; he must suffer for the good of the body politic'. Yet in the logic of Booth's social statistics, as illuminated by his 'personal observations', lay the outline of a different sort of society. Lord Beveridge said that it was Booth's London survey that gave him the ideas which animated that charter of human needs and rights, the Beveridge Report, of the Second World War.

And in the United States the bold example of Booth was a major force in starting a great series of social surveys, of Pittsburg—the first—in 1907, of Springfield, Illinois, in 1914, of Chicago in the 'twenties',[5] and beyond that, the revelatory cross-sectioning of 'Middletown' and 'Yankee City'.

'Wide Views in Unexpected Directions'

It was a sign of the times that the London Statistical Society had now quietly dropped its Latin motto—'To be Threshed out by Others'. As statistical science came into being, it was discovered that the threshing was indeed inseparable from the gathering in.

A powerful influence promoting this change was that of Adolphe Quetelet, a teacher of mathematics who became Belgium's Astronomer Royal, and who went on with missionary zeal to harness statistical techniques to what he called 'social physics', based on his discovery that such human phenomena as suicide, crime, marriage and so on showed a regularity in their occurrence down the years 'of the order of physical facts'. An acquaintance of the great French mathematicians, Poisson, Laplace and Fourier, Quetelet employed probability theory to demonstrate the distribution of human moral and physical traits around a mean, in curves conforming to mathematical laws.

Working on the heights of 100,000 newly called-up French conscripts, Quetelet announced confidently that 2,000 men had evaded conscription by somehow minimising their heights—since the recruits' heights failed to coincide with his prediction, calculated from 'the law of variation arising from accidental causes'. In a similar fashion Quetelet projected 'The Average Man', a statistical artefact which was to prove to be more central and compulsive than even Quetelet can have dreamed of.

Which was the more 'real'—the statistical abstraction, or the discrete, numerical 'facts' from which it was distilled?

In an age when faith in Science and 'Progress' was still hardly challenged such confusing questions were premature. The prophetic mantle of Adolphe Quetelet fell from his shoulders only to be picked up, firmly and promptly, in England by Francis Galton. It was strangely appropriate, because Galton, a Birmingham banker's son trained as a doctor, was, in a curious way, in the line of descent from that earlier prophet of number, Thomas Malthus. Galton was Charles Darwin's cousin, and, as Malthus's doom-laden progressions had suggested to Darwin the mechanism behind the 'Survival of the Fittest' in the Natural World, so, now, Galton made it his business to trace the workings of heredity in man, using, as he insisted, the *statistical method*.

But as, over thirty years, he traced the passing of traits in distinguished families from generation to generation, he became dissatisfied with the available criteria of human attributes. They had not, he complained, 'been submitted to measurement and number'—without which 'no branch of knowledge' could 'assume the dignity of a science'. He determined to remedy this deficiency.

At the International Health Exhibition, held in London from 1884-1885, Galton organised an 'anthropometric laboratory' where visitors of many 'races' were put through a battery of tests, mostly of his own devising. Numerical readings were registered for keenness of sight and hearing, colour sense, strength of pull and grip, force of blow, span of arms, height standing and height sitting, weight ...

Not only were the measurements meticulous, some of them were also mathematically correlated. Correlation coefficients were calculated showing the relation of pairs of attributes— of the height to length of the middle finger, of head length to head breadth, and so on. In the same way, in his work on heredity Galton prepared an index of correlation showing the deviation of individuals from the family norm, the share to be attributed to father and mother, even to Nature and Nurture.

It was, wrote Galton in his book, *Natural Inheritance* (1889), a novel idea 'full of interest of its own' and affording 'wide views in unexpected directions'. It was also to mark something of a watershed in the development of statistical science. In the words of Karl Pearson, one of the pioneers of modern mathematical statistics: 'Up to 1889, men of science had thought only in terms of causation; in future they were to admit another working category, that of correlation, and thus to open up to quantitative analysis wide fields of medical, psychological, and sociological research.'[6] In fact, added Pearson, the missing key that would unlock 'the treasure chest' had been found.

Thus when in 1892 Charles Booth received the first Gold Medal awarded by what had now become the Royal Statistical Society, 'the Numerical Method' stood on the verge of a breakthrough that would take it far from Booth's own simple 'figures given life by personal observation'—and, indeed, often far from the comprehension of all non-mathematicians. Malthus had clothed his moral philosophy in figures, but he had not abandoned words. Now, increasingly, the figures were self-sufficient, the proposition, the proof, the verdict, the Truth.

In the seventies the British economist, Stanley Jevons, declared economics to be 'a mathematical science' closely analogous 'to the science of Statistical Mechanics'.[7] Elaborate curve diagrams accompanied his Study of *Periodic Commercial Fluctuation*. His *Treatise on the Value of Gold* further sophisticated the technique of price index numbers. He was the first economist to use a semi-logarithmic grid in analysing economic data.

Walter Bagehot, the shrewd editor of *The Economist*, commented ominously: 'Mr Stanley Jevons and Mr Walras of Lausanne ... have worked out a "mathematical theory" of political economy; and anyone who thinks what is ordinarily taught in England objectionable because it is too little concrete in its methods and too unlike life and business had better try the new doctrine which he will find much worse on these points.'[8]

But everywhere the philosophers were in retreat, mathematicians, statisticians, 'sophisters and calculators' beyond Burke's wildest dreams were moving up. Inspired by Galton,

James McKeen Cattell had introduced 'mental' testing into America, Binet into France. Originating in the application of statistically designed experiments to agricultural crops, the new science of 'biometrics' in due course spawned a multiplying brood, psychometrics, sociometrics, econometrics as in one subject after another the Galtonian key to the treasure chest was turned.

There were those, however, who continued to be troubled by nagging doubts about what, thus opened, it might prove to contain. Not all of them were laymen. In 1932, Harold Westergaard, the Danish economist, and author of a history of statistics, commented that while the new mathematical methods 'gave statisticians easily handled formulae ... it cannot be denied that these formulae contained a danger, tempting as they might be to the too mechanical treatment of observations and, so to speak, to increasing the distance between the observer and the facts, whereas the old-fashioned methods had the advantage of keeping the observers themselves more in view, and therefore often made it easier to draw safe conclusions from the material.'[9]

These were to prove pregnant words.

Dr Kinsey Discovers the 'Human Male'

'The American people,' said General Francis Walker, organiser of the Tenth Federal Census, 'are intensely and passionately devoted to statistics.' It was a fact often remarked upon by travellers from Europe, and it was certainly not inappropriate that the United States should become the scene of that industrial revolution in statistics which was to complete the essential foundations of the Numbers Game as we know it today.

This industrialising process turned on two hinges. Firstly, Galton's key of correlation was no longer hand-turned, but spun by fast-running, electrically powered machinery. Secondly, the continuous flow of numerical verdicts and indicators that cascaded from the calculating machines was more or less automatically fed into a national projector—itself statistically monitored—the popular press, the radio, 'the mass media'.

In 1880 an American engineer named Herman Hollerith worked on that year's Federal Census, and learned at first-hand the problems of completing the tabulations in the ten years allowed by the Constitution. By the next census, of 1890, he had devised an electrically actuated punched-card tabulating machine, which could handle and sort 24,000 cards full of information in one hour. The dream of Charles Babbage of a calculating 'engine eating its own tail' had been brought a long step nearer fulfilment.

Indeed, an 'engine eating its own tail' of a somewhat different sort was then coming into existence in the shape of the circulation-monitored press, which both shaped its readers, and was in turn shaped by them. 'The Average Man' might indeed be a statistical figment, but in the headlines and the type columns beneath them, he took on a substance and sharpness of outline which in a society animated by an egalitarian philosophy could be compulsive.

In the United States through the twenties and thirties numerous surveys, radio ratings, straw votes, finally economic indicators and opinion polls underlined the natural attraction between such social indices and the headlines. But the statistical exercise which most vividly epitomised this affinity and its possibilities, signalling a new stage in the development of the Numbers Game, occurred soon after the Second World War. The subject was a survey of 'the sexual behaviour of the human male'; its director, Dr Alfred C. Kinsey.

The social scientist's classic aspiration to apply the rigorous methods of natural science to the study of man in society came very naturally to Dr Kinsey, who was himself, in fact, a natural scientist, an entomologist who had lately concluded a classificatory survey of a single species of gall-wasp, involving the capture and micrometer measurement of 150,000 variants. Certainly when, as Professor of Zoology at Indiana University, he found himself at a loss for adequate answers to the anxious personal sex questions of his students, he had no hesitation in seeking them through a survey to which he brought the basic approach and some of the techniques he had already proved in the gall-wasp inquiry.

The scientific method is global, but the Kinsey Reports could probably have happened only in America. Dr Kinsey brought to his gigantic, self-imposed task not only the dedication of the scientist, but the organising energies of the production engineer. In a nine-year operation he and his associates verbally addressed an average of 300 questions from a battery of 521 to 12,000 persons, aged three years to ninety, residing in every state of the Union. The personal sex histories, he tells us, covered five times more material than any previous sex survey. They embraced 163 representative groups (including 'hitch-hikers over a 3-year period'); contact men used in securing interviews ranged from bootleggers to 'persons in the Social Register'. A check system of retakes, comparison of partners' replies and so on monitored the accuracy of the accumulating data. IBM machines, fed with data on punch cards, randomised the 698 samples used in the survey, making them homogeneous for sex, race, marital status, age, rural or urban background, and religion.

The concepts of the production engineer equally dominate the questions. Sexual behaviour is seen in terms of 'outlets', 'frequencies', and 'techniques'. The 'outlets', all, of course, wholly physical, are six: masturbation, nocturnal emissions, petting to climax, heterosexual intercourse, homosexual outlets, animal contacts. And the finished product, the ultimate end, the summing up and *summum bonum*, automatically totalled as it comes off the Kinsey biological production line: the sum of orgasms, constituting the individual's output. 'Calculating from the age distribution of the total population and from the mean frequencies of the total outlet of each age group (Table 44), it develops that there are, on an average, 231 orgasms per week per hundred males between adolescence and old age.'

Such is life. There is little hint in this whole vast, extraordinary, joyless work that, for the human individual if not for the Naked Ape, these experiences may vary at all in quality and significance, or even that a complex relationship with another person, as distinct from a complementary mechanism, may be involved. (It seems entirely appropriate that the 'Male' and 'Female' surveys should be segregated in separate volumes, separated by five years.) For Dr Kinsey, the scientist, such human elements which can be neither measured nor counted are apt to smell of mysticism or, worse, moral prejudice. The personal impressions and insights on the spot which Charles Booth valued as bringing correcting 'colour' to his bleak statistics, for Kinsey are apt to smack of what he witheringly calls 'barber-shop techniques'. He concedes that 'sexual responses involve emotional changes,' but then goes on to ignore them. 'Kinsey,' commented the American scientist, Dr Iago Galdston, 'is cognizant of the *id*, but he seems to look upon the *superego* as the malignant artefact of a senile and malevolent society.'[10]

Indeed, Kinsey does not merely recognise the libido, he enthrones it, always provided that it is manifest in the form of *measurable* 'outlets'. He makes, for instance, an obviously heartfelt protest against the application of the term 'impotence' to premature ejaculation, 'however inconvenient from the standpoint of the wife in the relationship'. 'It would,' he comments, 'be difficult to find another situation in which an individual, who was quick and intense in his responses, was labelled anything but superior ...' and he reminds us that we should 'benefit from a more general understanding of mammalian behaviour', and that ten to twenty seconds is all that is ordinarily required by chimpanzees.

In many ways, Kinsey seems to prefigure—and even at times caricature—some of the characteristic features of the Numbers Game of today. 'Scientific' objectivity places at the centre whatever has the supreme virtue of being measurable; what cannot be measured does not exist; there is the preoccupation with animal behaviour (animals cannot talk back); and there is the tremendous reliance on mechanical correlation, and the sheer, numbing weight of 'semi-automated' statistical 'results'. Hollerith punch cards, and now the IBM machines, had made it possible to pour out correlations by the barrowload: the Kinsey Report on the Human Male, with its 160 tables and 170-odd graphs, floats on a vast, flaccid sea of them.

In the ritual reservation that soon becomes so familiar to any reader of social science research findings, Kinsey himself points out that 'simple two-way correlations at best show a relation, but not necessarily a causal relation'. (Nor, for that matter, frequently, do unsimple correlations.) But as the successive correlations thud down, one after the other, the lay mind at least is anaesthetised, and the suggestion of cause-and-effect becomes irresistible. Nor indeed does Kinsey, the scientist, himself appear to resist it particularly strenuously, certainly not when biological drives appear to be hampered by 'inhibitions' (alternatively ascribable to superstition or civilisation). Thus a correlation of figures for pre-marital intercourse and 'sexual effectiveness in marriage' measured by 'the occurrence of orgasm in marital coitus' ends as an implicit argument for pre-marital sex. Or, as Gertrude Stein might have more economically put it: an orgasm is an orgasm is an orgasm. Again, it is the sort of exhilarating round trip that becomes familiar to regular players of the social science Numbers Game, and, proof-wise, is as unassailable as 666, the Number of the Beast.

But perhaps Dr Kinsey's most notable contribution to the development of our world of the socio-statistical artefact and its orbiting self-fulfilling prophecies derives from the manner of the Kinsey Reports' publication. Earlier sex research had been filtered through the specialist press where its limitations could be appreciated, and the particular focus in due course blended into the larger picture. (Somewhat earlier, portions of Krafft-Ebing's works had been printed in Latin.) The Kinsey Message—and that there *was* a message seemed implicit—was revealed to a world press rendered expectant and on tenterhooks by much skilful advance publicity.

Never had the natural attraction between statistics and popular press headlines been more vividly demonstrated. '87 per cent of American 30-year-olds Have Pre-marital Intercourse'—'70 per cent of American Men Visit Prostitutes'—'37 per cent of American Males in Homosexual Activity'—'97 per cent of American Men Have Committed Sex Crimes.' It had impact; it was crisp and precise; it packed a wonderful amount into a three-column headline; and, most of all, it couldn't be faulted: it was authoritative, objective, *scientific*.

Despite its high price, many tables and leaden prose, *Sexual Behaviour in the Human Male* for a while outsold *Forever Amber*. Six printings vanished in January and February 1948. Never before in history had a scientific book, not even Darwin's *Origin of Species*, enjoyed such a sale. And the 'Human Male' volume served as a protracted trailer, building up the suspense before 'K-Day', as the newspapers were now calling the publication date of the Report on the Human Female which burst upon the world's press five years later, to receive coverage of an enthusiasm and detail hardly seen since Lindberg flew the Atlantic.

As with Charles Booth's unveiling of poverty, the secret lay in the detached authority, the starkness of the figures. Others, it is true, had undressed sex; but Freud had substituted other, yet more bizarre, garments for those he removed; Havelock Ellis, although specific enough, had wreathed his biology in poetic idealism; and Baron von Krafft-Ebing had, avowedly, dealt in morbidity. The Kinsey reports cut through the curious blend of Viennese nebulosity and Sunday school sin-and-giggle that had survived to reveal 'the facts of life' in all their arithmetical prosaicness and absurdity. Whereas Freud had long ago written of infant sexuality, Kinsey enumerated it, and pointed to a peak of 'total outlet' in the mid-teens, with variagated curves of 'activities' declining steadily thence to old age. Whereas Krafft-Ebing pointed to secret depths, Kinsey plumbed them and, slide-rule in hand, issued his verdict: 'some types of sexual behaviour labelled abnormal or perversions in textbooks prove upon statistical examination to occur in as many as 30 or 60 or 75 per cent of certain populations'.

Here the statistical artefact exerts its influence not merely politically, or socially, but also acts on the most complex, innermost, yet highly suggestive, areas of the human consciousness. It was often said in praise of Dr Kinsey's work that his much published statistics had removed a great burden of needless and morbid anxiety. To the questions of his students and clients 'Am I normal?' Dr Kinsey was now able to return a resoundingly reassuring reply, backed by all the authority of statistical science. In a population as large and varied as that of the United States the punched cards flashing through the machines showed that almost anything was indeed normal. Kinsey describes how, as a taxonomist, he had found winglengths of insects varying between 10 and 180 micrometer units, a 'difference of 18 times' which probably represented the extreme of linear variation of any species of adult plant or animal. However, in his survey of the sexual behaviour of the human species, he had encountered 'one male (a scholarly and skilled lawyer) who had averaged over 30 ejaculations per week for thirty years' and others with a 'frequency' of once in thirty years. Dr Kinsey adds:

> ... attempts to recognise such states as nymphomania and satyriasis as discrete entities can, in any objective analysis, refer to nothing more than a position on a curve which is continuous.

'Normal or abnormal,' he concludes, 'one sometimes suspects are terms which a particular author employs with reference to his own position on the curve.'

Nevertheless, however it may be for those admirable chimpanzees, a position astride a continuous curve does not appear to many human beings as secure as no doubt it ought to be. The groping quest for models and standards, the apprehensive questions about 'normality', remain normal for the young. And, in fact, Dr Kinsey does not disdain to respond, at least by implication, as 'averages' and percentages pour from the calculating machines. It's all there in the tables—Mean Frequency, Petting to Climax, for three age groups, three education levels, three religions (active and inactive branches); or Intercourse with Prostitutes, Males, aged 8 to 45, percentages; or Six Outlets, as experienced in seven occupational classes ... 'The Kinsey Report,' commented Dr Millicent McIntosh, President of Barnard College, Columbia University, 'uses all the techniques to which Americans are especially vulnerable ... If the Kinsey Report announces that 81 per cent of females have done petting by the age of 18, the girl who is pressed by the boy to go further ... feels herself trapped by these statistics ... a counsellor in a university stated that many boys felt they were not actually virile if they could not keep up with the statistics Dr Kinsey presents for their age group.'[10] At the universities of California, Indiana and Minnesota, surveys among students showed that the Report had, in fact, changed attitudes.*

And here we come to the curious propensity of statistics, intended as means to a larger end, to become, under the con-

---

* In a cross-America questionnaire survey of college students in the late sixties, about twenty years after Kinsey blazed the trail, Vance Packard reported an almost 60 per cent increase in 21-year-old girl students owning to intercourse—a score of 67 per cent against Kinsey's 43 per cent. Perhaps more significantly post-Kinsey, however, is the remark of an American student, quoted by Dr Spock, that he, his male friends, and their girl friends

ditions of our contemporary world, ends in themselves. It is a distortion and inflation which is a central feature of the Numbers Game—a peremptory and mechanical foreshortening of life and experience which may emerge as this book proceeds, and become one of its central themes. And the curious thing is that often enough the statistics which bring down the guillotine are admittedly figments of dubious abstraction. Dr Kinsey himself pointed out in his reports that his 'frequencies' are, in fact, no more than approximations, and that his statistical averages 'suggest a regularity in the occurrence of activities which do not actually occur with any regularity'. His professional critics went on to indicate serious flaws in his sampling methods arising notably from his reliance on organised groups. Yet once committed to public print, objectified in innumerable headlines or in, say, the women's magazines, many of which treated Kinsey's female report as a sort of vade-mecum (the magazine *Pageant* printed a personal check-list for readers to mark off their ratings), the qualified figures stood forth like beacon lights, lucid, commanding, scientific—the *specifications*.

Common experience, social history and recent research all suggest, on the contrary, that human sexual—as other—behaviour is varied and reasonably malleable, adapting to circumstances. 'The adolescent boy's erotic urges,' writes Professor Frank A. Beach, a veteran American researcher in the field, 'stem more from socio-cultural factors than from those of a strictly psychological nature.'[11]

But life at mid-century was increasingly imitating not Art, but Science, and for more and more people science meant, in effect, the statistics of social science (particularly when geared up with other statistics, as we shall see in the next chapter).

By the inner nature of their calling, statisticians in the social field, their minds mathematically structured, are discoverers of regularities or 'laws', seekers of models and systems. Even with

---

now tended to 'spend a lot of their dating hours discussing the pros and cons of various stages of physical intimacy, as if the couple were taking part in a university seminar in the physiology and psychology of sex'—the poetry of the booming sex questionnaire and technology industry. (*Decent and Indecent: Our Personal and Political Behaviour*: Dr Benjamin Spock, 1969.)

his relatively simple mathematics, Malthus presented human life as an equilibrium precariously maintained against disease and starvation. Kinsey equally centred on the biological drive, rendered 'the passion between the sexes'—Malthus' great 'constant'—into statistical forms. But in his, Kinsey's—equally 'objective'—model, Nature, of course, was not to be feared but to be embraced. But whether the 'natural law' you preferred was that of religion or biology, whether you saw 'temptations' to what Malthus called, comprehensively, Vice, or inhibitions ignorantly interfering with what Kinsey called Erotic Response, whether your source of authority was God or Science or both, life viewed through the statistical binoculars was apt to look predetermined.

Neither Malthus, nor Booth, nor Kinsey were original in their subject matter. Their distinction lay in their creation of powerful statistical models or artefacts. In this indeed they were true and massive innovators, makers of history in both the literal and general sense.

The grim model of Malthus left its stamp on society for three generations. Booth's Poverty Line led to the comprehensive Welfare State. As for Dr Kinsey, we are still tossing and turning in the turbulent and murky wake of his statistical craft, But there can be little doubt that the impact of his mechanised accountancy has penetrated much further than that of any earlier, merely literary, evangelist.

Further—and faster. Looking back now from what has no doubt exaggeratedly, but excusably, been called 'a decade of orgasmic preoccupation', it is hard to recall that as late as 1950 Hollywood film producers were being 'regularly cautioned that a married man and woman should not'—in the unique language of the Hays Office—'be overly eager to exercise their marital privileges' while any appearance of an unwedded couple on a bed was totally taboo.

In America, wrote Max Lerner, 'Kinsey's impact was that of a kind of Guilt Killer,'[12] and this in a country with so rooted a Puritan tradition was no small thing. Equally in England, the twenty years after Kinsey brought a radical transformation extending across a wide range of critical socio-moral laws. Homosexual conduct between consenting adults in private was taken out of the sphere of the criminal law; divorce by consent on the breakdown of marriage was at long last conceded; abortion law was liberalised; in July 1969, the pre-censorship and licensing of plays by the Lord Chamberlain, which had been in existence in England since the beginning of the eighteenth century, was suddenly abolished.

But if the Kinseyian revelation could demolish obsolescent standards and clear our minds of much old cant, it could not create new values to replace them, and was to develop a voguish new cant of its own. In some ways Kinsey merely preserved the Victorian dichotomy of Sacred and Profane Love, reversing positions, and sanctifying or certificating the Profane with statistics. Kinseyian statistical 'normality' was a strange construct in which sexual 'activity' proceeded *in vacuo*, divorced from emotion, personal character, family, work, psychology, ideas ...

Malthus, who had a wife and three children, oddly enough makes the point vividly in his first *Essay on Population*:

> It is the symmetry of temper, the vivacity, the voluptuous softness of temper, the affectionate kindness of feelings, the imagination and the wit of a woman that excites the passion of love, and not the mere distinction of being female. Urged by the passion of love, men have been driven to acts highly prejudicial to the interests of society, but probably they would have found no difficulty in resisting the temptation had it appeared in the form of a woman with no attractions whatever but her sex. To strip sensual pleasures of all their adjuncts ... is to deprive the magnet of some of its most essential causes of attraction ...

It was an impoverishment which, however, over large areas now appeared to be increasingly insisted upon. In England, following the liberalisation of the new Obscene Publications Act (1959), film censorship was greatly relaxed; in 1966 in the United States the 'Hays' Code at last came to an end, and a new code proclaimed among its objects the 'expansion of creative freedom'.* Freedom indeed expanded spectacularly in one direction at least, but creativity was more in doubt. Frequently cloying and unreal in the forties, film love by the late sixties

* In 1952 the US Supreme Court reversed its 1915 ruling that motion pictures, being 'a business pure and simple', were not covered by the First and Fourteenth Amendments, guaranteeing freedom of speech and the Press. Censors were therefore in any case on more delicate ground than before.

had taken on a routine physical literalness, a depressingly repetitive determination to leave so little to the imagination that a film-goer might be excused for gaining the impression that Dr Kinsey or one of his acolytes was somewhere just off-shot with a coded record sheet, adding his conscientious tabulations to those of the Box Office.

The bookstalls reflected the same preoccupations with a possibly even greater monotony. A reader wrote to the London *Times* to say that, in 1970, in a quarter-hour's search for reading matter on a main London railway bookstall he had been able to find nothing but sex, violence, and the *Forsyte Saga*. (And after all, in the Kinseyian accountancy, pornography, provided only it produced physical effects—which the US Commission on Obscenity and Pornography stood ready to check with the 'penile plethsmograph'—was an 'outlet' as aggregative as any other.)

In the autumn of 1970 no fewer than half the first ten nonfiction books on the *New York Times* Bestseller List were about sex. Indeed, proceeding with the post-Kinsey process of 'killing guilt', the United States appeared to be moving towards its own *Kama Sutra* or Ovid, although somehow what emerged was not lubricity or the 'Art of Love', but the clinicism or technology of sex in do-it-yourself manuals as from the pen of a conscientious, but facetious, plumber.

As late as 1960 in England the award of an 'X' (adults only) category to a film was still considered by the major circuits to detract from its commercial possibilities. Ten years later endless 'sexploitation' films, bringing automatic increases in takings, were being widely seen as the road to financial salvation.

But for the driving power behind such changes, so powerfully affecting the cultural environment and social values, we must look beyond the statistical artefacts of social science, influential as these can be, to the point where they mesh—as this book suggests they do mesh—with other monitors and other forms of accountancy.

3

AMERICAN SYNCROMESH:
THE 'CONTINUOUS AUDIT'

Newspapers, I think, resemble fashionable ladies of the West
End in that they are more concerned with their figures than
their morals.
> Colonel the Hon. E. F. Lawson, former General
> Manager of the *Daily Telegraph* to London Uni-
> versity journalism students.[1]

A joke of the period represented a motion picture magnate as
saying 'Two and two make four, four and four make eight,
and sex and sex make millions.'
> Mark Sullivan, *Our Times*, Vol. IV, 1909-14.

Dr Kinsey's 'ratings' offered merely one particularly vivid illustration of the way in which in modern society our consciousness, idea of ourselves, life-style, even our morals are increasingly shaped and monitored by statistical findings which are
turned into powerful social artefacts by continuous projection
from the gigantic two-way mirror of the media.

It is a complicated, multi-geared, now semi-automated process in which the design and economic logic of the machinery
of projection itself plays a crucial, if often hidden, part. For this
is a mirror which refracts as well as reflects and, on its 'other',
operating, side, its focus is adjusted and its projection machinery tuned by its own interlocking, statistical monitors. Half a
century before the electronic computer instituted the 'age of
automation', the machinery of the media was already developing its own systems of 'feed-back', numerically based on the
movements of daily net sales, box-office totals, audience ratings,
percentage of AB readers and the rest.

Brought today to a high pitch of mechanical sophistication
and minute-to-minute continuity, this monitoring mechanism
lies near the heart of the complex and towering structure of
consumerism, supporting and guiding the dynamics of marketing and salesmanship which raised it. Thus ordered, the present

arrangement and machinery of the media takes on an air of inevitability, as though it were a fact of Nature rather than the
work of men.

For perhaps the most notable feature of these mechanical
or semi-automated monitoring processes which so characterise
our times lies in the success with which they mask where
control lies, and what motivates it. They offer, instead, the
instant alibi of democracy, vast majorities certified by the
meters. It is for this reason that it is so necessary to look behind
the gigantic and hypnotic mirror in which our portrait appears
outlined in numbers, to see how it was put together, what is its
inner logic, or how and why and by whom its monitoring
controls were programmed.

Figures Make News, News Makes Figures.

On any reasonable count this gigantic social mirror, with
its built-in, semi-automated controls, must rank with the
assembly line and the cheap, mass-produced automobile as one
of the most potent contributions of the New World to the Old.
And, as with the cheap, mass-produced car its, in the main,
American origins are certainly no accident. All nations, like all
men, need from time to time to confirm their identity; America,
however, had to create hers, and create it over and over again.
The first mass society in the world, built on an influx of immigrants from the four corners of the globe, she needed,
urgently, to see and to measure herself as she changed and grew.

Whereas in Britain a cheap and popular Press was deliberately
barred by a high Stamp Duty until as late as 1855, in the United
States the freedom of the Press was specifically guaranteed by
the First Amendment to the Constitution in 1791, while the
Separation of Powers meant that much discussion which in
Britain took place within that good club, the House of Com-
mons, would, in America, enliven the pages of the newspapers.
In such a vast, empty land, with its moving frontier, the news-
paper Press had from early days a very special function. 'The
printing press,' wrote a newspaper pioneer of the Middle West,
Captain Henry Kirby, 'preceded all the usual agencies of society.
It did not wait for the rudimentary clutter of things to be com-

posed and organised. The spirit of adventure thrust it forward ahead of the calaboose, the post office, the school, the church, and made it a symbol of conquest. Thus the theory of publicity was emphasised as a factor in the westward march of the American people and their institutions ...'[2]

But as the country was linked coast to coast by rail in 1869, and the cities boomed and the immigrants poured in, journalism of this grass-roots, face-to-face sort was transcended. Now projecting and magnifying, the mirror grew more complex; human individuals gave place to 'human interest stories', people to percentages and averages. The 'theory of publicity' was systematised: the newspaper branch of the Numbers Game inaugurated.

One September morning in 1884 the citizens of New York were startled by the boom of guns in City Hall Park. It was a salute of a hundred guns to mark the attainment of a circulation of 100,000 copies daily by the *New York World*, more than six times the sales of the paper when Joseph Pulitzer, moving up from St Louis, had bought it only a year earlier. Superficially, the secret of this historic feat of levitation—the prelude to many others—was simple; in fact it was many-sided, demanding great editorial flair. It included ultra-sensational headlines, bold illustration, good domestic news-coverage, high-minded editorials, simple writing, and, above all, resounding stunts and crusades, incessant self-promotion.

By fixing advertising rates initially low, Pulitzer ensured a large paper; by gearing them to certified net sales figures he ensured that the advertising pages soon bulked as large economically as they did physically. Although his first issue of the *World* in 1883 had declared that henceforth the paper was to be 'dedicated to the cause of the People, rather than that of purse-proud potentates', he was now speedily demonstrating that one of the fastest ways to qualify as a purse-potentate could be to own a newspaper. In particular, the *World*'s Sunday edition, prodigiously swollen with advertising, appeared to be a gold-mine.

By 1886 the *World*'s circulation had reached 250,000, and a contemporary was noting that its success 'had affected the character of the entire daily press of the country'. At these newly inflated levels Pulitzer had imposed a new hard logic on newspaper economics. As the key to advertising revenue—now more critically important than ever before—circulation figures came under a continuous spotlight. Newspaper production often became a desperately run race, or gladiatoral contest, in which number was all. Those falling behind, whatever their merit, went to the wall. It was a process whose mindless imperatives determined not only the character of the papers, but sometimes the shape and selection of the news itself.

Again, this aspect of the Numbers Game received a classic demonstration very early in the day in America. In October 1895 William Randolph Hearst, who had remodelled his father's *San Francisco Examiner* on the Pulitzer pattern, came over to the East Coast, bought the New York *Morning Journal*, and proceeded to try to out-stunt, out-promote and out-sell the master in his own backyard. He plastered the street-cars, the elevated railway, mailed cents to registered electors to enable them to buy the *Journal*. He hired brass bands and set up 'free coffee' waggons, and bribed away the entire staff of the *World*'s Sunday edition, together with the artist who drew the *World*'s famous Yellow Kid cartoons—so called because the kid's apron was printed yellow. For a time two rival Yellow Kids touted their horrendous charms from the hoardings.

Thus, in more senses than one, 'Yellow Journalism' was born of the Numbers Game. And not only yellow journalism.

In 1897 the *Journal*'s sales figures passed the *World*'s. Pulitzer's famous editor, Arthur Brisbane, had deserted him for Hearst who offered a salary which increased by one dollar for every thousand additional copies sold. The timing of the deal was excellent: the greatest circulation builder of them all was signalled.

The Spanish-American war, which broke out over Cuba and the loss of the *Maine*, was to a remarkable, unquantifiable, extent a 'circulation war', a by-product of the struggle between Hearst and Pulitzer. And although this is certainly an extreme case it may tell us something about the potential of newspapers shaped by feed-back from their circulation graphs. Hearst's *Journal* boomed on a continuous diet of lurid, unchecked, and boldly splashed atrocity stories in which the Spanish authorities figured as 'butchers', rapists, torturing tyrants ad infinitum. It blazoned endless 'Insults to the American Flag'; staged heroic rescue expeditions. And although Pulitzer and the *World* were normally liberal, he simply could not afford—as he himself candidly admitted—to disregard such wonderful circulation stimulants. The *World* went after the *Journal* with atrocity for atrocity, insult for insult, call for action for call for action.

Both papers passed the million mark.

What now? President McKinley was in favour of non-intervention. After the mysterious explosion of the *Maine* in Havana harbour, Spain was ready to make every possible concession. But only the final step of war could hold the circulations of the *Journal* and *World* at their new heights.

The 'mirror' had never been so brilliant. On 2 May 1898, the *Journal's* front page was entirely filled by poster-sized type: WAR—MANILA IS OURS. And in a triumphant panel at the top of the page: 'Saturday's Circulation: 1,408,203.'[3]

On the other side of the Atlantic the opening stages of the media Numbers Game unfolded a decade or so behind events in America, and with sedater pace and style. Though less sensational and demagogic than Pulitzer's *World*, Alfred Harmsworth's halfpenny *Daily Mail*, launched in 1896, applied a basically similar formula. Like Pulitzer, Harmsworth was highly conscious of the dawning world of consumer business and advertising. He gave the *Mail* a magazine page, two columns for women, a column for the small investor. He instructed his reporters to write for tomorrow's £1,000-a-year-man'. He much admired the American departmental store impresario, Gordon Selfridge, late of Marshall Fields, Chicago, who had opened a vast pillared emporium in Oxford Street, and, most promisingly, believed in liberal use of displayed advertising. He told his editors to 'print a good story every day about the big stores'. One of Selfridge's directors later conceded that 'Lord Northcliffe (Alfred Harmsworth) had helped to sell the shopping idea to the general public.'[4]

Yet Northcliffe's most important innovation was made even before the *Daily Mail's* birth. In 1892, his weekly *Answers* printed the first British Net Sales Certificate, 'issued by a Chartered Accountant'. In Britain the Audit Bureau of Circulations was not finally established until 1932, eighteen years after one had begun to operate in America. But more than ten years earlier Northcliffe introduced advertisers to that precise and deadly statistic, 'cost per thousand readers per single column inch'. The *Daily Mail* reached its first million during the Boer war; but by the thirties the all-important figure had become two millions, first reached by the *Daily Herald* in 1933, with the aid of cascades of gifts, cut-price sets of Dickens, and an army of canvassers. The cost—up to £1 for each new, and probably transitory, reader—was ruinous, but as one of the Labour *Herald*'s former editors, the late Lord Francis Williams, has explained: 'unless it could flourish a net sales certificate of the order of 2,000,000 in front of advertisers, it could not hope to pay its way on the level at which it had been forced to establish itself to get into the game at all.'[5]*

The representation of a sector of opinion, the steady performance of a useful democratic function, was no longer enough: position in the race was crucial. By 1937 the *Herald* had lost its lead to the *Express*, and—in the dispiriting logic of this Numbers Game—nothing failing like failure, had begun the long descent which, despite a million sale, was to end in the abandonment of the politically historic title in 1961. While from 1950 two papers, the *Express* and the *Mirror*, enjoyed a combined circulation equal to all the seven other national dailies put together, the squeeze against the middle-size papers representing a diversity of views and positions (which the Royal Commission on the Press of 1949 had particularly wished to encourage) was now continuous. The Liberal *News Chronicle*, again representative of an important body of opinion and of a certain very English temper, shut down in 1960, allegedly economically unviable despite a circulation still at the level of 1,200,000.†

The heart of a newspaper was now less and less likely to be found in its character or philosophy, spirit or tradition. It lived and breathed by its circulation graph, an automatic monitor of great, if silent, power, shaping not only the Press, but the news —and the very concept of news. The greater the escalation of circulations, the greater the funds at risk, the more intimidating

* An almost irrational importance was attached to successive large round numbers—one million, two million, and so on. Thus William Hardcastle, a former editor of the *Daily Mail*, during its postwar travails, writes of the 'magic figure' of two million. The need to keep the paper above this mark was 'considered paramount'. (*Campaign* article, 12 March 1971.)

† For further facets of the newspaper Numbers Game operating against such papers, see later section in this chapter.

any downward fluctuation. For it would be noted, not only by editors and managements, but by advertising agencies always alert for signs of failing strength, and perennially reluctant to put clients' money into a loser.[*]

Indeed, the approaches of journalism and advertising now overlapped: in popular papers at least it was necessary not merely to report, but to package and market, to 'sell' the reader. A 'good story' might, or might not, be a significant story; but it must startle, catch the imagination, or appeal to a small number of basic emotions. In such circumstances, much 'news' is bred incestuously. Hamilton Fyfe, an old Northcliffe editor, long ago pointed out that 'if a mood seems likely to be popular, every popular paper will inflame it, so as not to let its rival get ahead.'[6] Certainly, under the relentless pressures of the Numbers Game, competition between newspapers may make not for diversity and 'freedom', but—as the early instance of the two Yellow Kids suggested—for a paralysed copycattery, more concerned to 'spoil' the rival's efforts than to develop creative—and therefore risky—ideas for oneself.

Certainly, any effective newspaper must closely reflect its readers and its society. But the newspapers a democracy needs will do more than reflect; they will have, as C. P. Scott put it, a 'soul of their own', the integrity to take a steady view, and to express this clearly even if it conflicts strongly with that of the overwhelming majority. As late as 1915 even Northcliffe himself could court great unpopularity in attacking the people's hero, Kitchener, for failing to provide sufficient HE shells on the Western front. The *Daily Mail* was bombarded with cancellations, advertisers withdrew and circulation fell 15 per cent. The question now is whether at this stage of the newspaper Numbers Game and at today's near-saturation levels of circulation, any large newspaper could persist in taking a stand against strongly held majority opinion, once the ever-present monitors had turned decisively down.

'A newspaper,' ruled Mr Hugh Cudlipp, then presiding genius of the very successful 4m sale London *Daily Mirror*, 'may successfully accelerate, but never reverse, the popular attitude ...'[7] Certainly, in this situation it might be unprofessional to try. For now the more urgent imperatives of the Stock Market and the tell-tale, ever-busy indices of business accountancy

heavily underscored each fluctuation of the circulation graph.*
The *Daily Mirror*, self-appointed tribune of the people, was
owned by the gigantic International Publishing Corporation, a
vast amalgamation of printing, paper-making, publishing and
magazine interests, capitalised at £90m, with 54 per cent of its
shares held, in 1968, by City institutions, seasoned number-
watchers, not unaware that the Truth which is said to make
some free may make others speedily bankrupt, and unlikely to
respond to the vibrant appeal of Ebenezer Elliott, the Corn
Law poet:

> 'O, pallid want, O, labour stark,
> Behold we bring the second ark,
> The Press! The Press! The Press!'

What Price the Stars?

Although today the hegemony of Hollywood is long over,
and the film has made a break for freedom, historically the rule
of number and the systematisation of feed-back from the
numerical monitors might not have attained the authority it
did, had not the mind-stunning totals of 'Box Office' at a
critical moment lent powerful reinforcement to the new absolutes of the Net Sales Certificate. In Britain the class structure
of society still limited the field of the rising, mass-sale newspaper ('written,' as Lord Salisbury had sneered of the *Daily
Mail*, 'by office-boys for office-boys'); in America the sheer
vastness of the land defeated even Mr Pulitzer, and Mr Hearst's
headlines and newspapers remained merely regional.† But, well
before radio, the movies had begun to break the bonds both of
geography and social structure, giving a high gloss and a
larger than life image to the building mirror of Narcissus, im-
porting that continuous element of fantasy which made the
levitations of the Numbers Game possible.

* The point was vividly underlined in 1968 when the chairman and
architect of the IPC, Lord Northcliffe's nephew, Cecil Harmsworth King, who
had been taking a strong political line in the *Mirror*, finally calling for the
Prime Minister's resignation, was summarily ousted by the vote of his own
Board. With IPC's profits falling steeply, and the dividend about to take a
savage cut, such political excursions had become a luxury the company
could not afford.
† Despite the establishment of McClure's syndicating agency in 1884 and
of the Associated Press wire service in 1892.

Furthermore, the mechanics of the motion pictures were not inhibited, at first at least, by any ideas of sacred mission or élitist standards. The movies were a popular art and a showman's business from the start, dedicated to simple amusement and simple money-making. Fox, Zukor, Laemmle, Loew, all the great movie tycoons, began by making money in nickelodeons. Their motivations remained frankly commercial. In setting up a 'national film exchange' in 1913, William Fox, a former garment worker, called it simply 'The Box Office Attractions Company'. But, already, four years earlier, there had been a fascinating glimpse of the way things were moving. The 30 September 1909, issue of the London film trade paper, the *Bioscope*, contained an account of an 'Italian gentleman' who had been introducing into audiences an invention called a 'psephograph'— 'something like a penny-in-the-slot machine, with four dials marked "good", "bad", "indifferent" and "total". By placing a metal disc into one of the three slots one's opinion may be automatically registered.'

And from the early days a geographical accident ensured the absolute dominion of the rulings of the Box Office. While the actual film-makers were drawn to California by their need for the sun, the proprietors and controlling money men continued to operate for the most part from New York. As the industry grew more complex and highly capitalised, the 'Front Offices' developed elaborate controlling hierarchies for whom the studios were 'the plant' and the pictures 'the product'.

The banker producers, wrote the historian of the American cinema, Lewis Jacobs, analysed projected movies 'for the following selling points: (1) "Names"—that is, stars. (2) "Production Values"—elaborate sets, big crowds, and other proofs of great expense. (3) "Story Value"—the huge prices paid for the original and its great reputation as a novel or a play. (4) "Picture Sense"—a conglomeration of all these items. (5) "Box Office Appeal"—plenty of all the standard values which had proved successful in years past.'[8]

First tried out around 1910, the 'star system' was designed to ensure a guaranteed return on investment on much the same principle as the promotion of national branded goods. The stars were exercises in financial statistics, their much publicised fabulous salaries functions of their Box Office pulling power (or 'marquee value') which was closely watched and reported on at frequent intervals. In interviews with filmgoers the Motion Picture Research Bureau compiled star popularity ratings, computed from a series of replies evaluated from 100 to zero—'I dislike Star X very much.' Audience Research Inc. broke down the ratings according to sex, age, income and frequency of attendance, 'so that a film producer can compose a cast which will appeal to many different sections.'[9]

The stars brought a new dimension both to the mass mirror and to its numerical controls. For the stars were 'ordinary Americans' who had 'made it', and by 1920 there were a score of fan magazines, full of the opulent mansions, swimming pools, and lordly lives of the stars. 'All ideas of duty, justice, love, right, wrong, happiness, honour, beauty, all ideas regarding goals of life ... are ideas planted by the movies,' reported Maurice Maeterlinck from the United States in 1921.

With admissions to America's cinemas rising to 100m a week in the thirties, when the studios were churning out 500 films a year, departmental stores all over the country received a flow of style reports from over fifteen Hollywood spotting agencies. Between three and four thousand Hollywood-based correspondents sent out an estimated 100,000 words of film news and features every day, and in most of them values and priorities were determined by statistics of one sort or the other.

The Numbers Game had found the perfect soil to bring it to full flower. 'A producer,' reported the anthropologist, Dr Hortense Powdermaker, 'would ask for a $750 a week writer or a $2,000 dollar writer.'[10] It was a truism that the worth of a picture depended upon the amount of money expended. The president of Paramount ruled that any picture costing less than one million dollars was a 'B'—or second-rate—picture.

As costs escalated in conformity with this criterion, each successive triumph required a larger and larger audience—and thus more and more expenditure on promotion—in order to break even. 'Pre-testing' of stories, vetting by the Hays Office, and the devices of feed-back were felt to be imperative to ensure that no possible customer was offended. In 1936, the chairman of the British Board of Film Censors, Lord Tyrell, announced that 'the cinema needs continued repression of controversy in order to stave off disaster.'[11]

Programming the 'Standard of Living'

If Hollywood immensely stepped up the projective power of the mirror in which America—and, after her, much of the Western world—saw herself, it remained for other media and their monitors to sharpen the day-to-day focusing.

An outstanding contribution was made by the parallel development of that uniquely American institution, the national mass-sale 'slick' magazine which owed its very birth to a sophisticated use of potential demonstrated in the newspaper branch of the Numbers Game. A low-priced—5 or 10 cent—'luxury' product was heavily promoted on the national scale, and the large sales attained used to attract a volume of advertising which in turn made the low—bargain—price possible. The formula worked brilliantly because, historically, its pioneers were enjoying a once-for-all opportunity. Spanning the vast distances of the North American continent as no daily newspapers could hope to do, the magazines opened up a vast national consumer market. They transferred power from the local retailers to manufacturers of branded goods, and en route generated fabulous profits for both advertisers and publishers.

A classic case was that of the *Saturday Evening Post* which Cyrus H. K. Curtis picked up derelict for $1,000 in 1899, and shortly afterwards announced was 'to be pushed into a circulation exceeding that of any weekly in the United States'. Curtis had started out as an advertising salesman in Boston, and his faith in the power of advertising was vast. He poured enormous sums into the *Post's* promotion and was spectacularly rewarded. By 1925 a single issue had an advertising revenue of $1,000,000 against a sales revenue of a mere $80,000.

When he relaunched the *Saturday Evening Post*, Curtis already had behind him the triumph of *The Ladies Home Journal*, which he had developed out of a rudimentary women's supplement, run by his wife, in a small farm weekly issued from Philadelphia. By 1895 it had reached the then prodigious sale of three-quarters of a million. By 1948 advertising revenue for a single issue reached $2,677,260, although by this time there were twelve American magazines with sales in excess of three million per issue.

Although they were powerfully to affect the shape of things elsewhere, these developments were in fact a response to a unique situation. Not only did they occur when a national consumer market was for the first time becoming practicable in the vastness of America, but also when millions of immigrants were groping their way towards a new style of life in their adopted country. Pulitzer, a German-speaking immigrant from Hungary, had championed their interests in his New York *World*. Now in *The Ladies Home Journal*, Curtis' famous editor Edward Bok who, like Pulitzer, had arrived in the country —in his case from Holland—unable to speak the language, carried the course in Americanisation a long stage further, detailing and illustrating an American—if not perhaps yet 'the' American—way of life. He provided model house plans, schemes of furnishing and decorating favoured by the 'Best People', portfolios of reproductions of the '250 Best Pictures'.

'Reading of magazine articles and advertisements is every year making customs and fashions uniform over the country as a whole,' rejoiced an advertising agent, Frank Presbery, in the twenties. The coupling of articles and advertisements in this way excited little comment. In the Curtis publications page-size was increased so as to offer better advertising impact, and the advertising pages interleaved with the editorial throughout. Edward Bok claimed to have originated in *The Ladies Home Journal* the practice of 'tailing'—extending the latter part of stories in single column through the magazine, so that the reader, as she followed the tale of romance or adventure in some nice American home, was simultaneously presented with an illustrated and priced catalogue of its appurtenances.

Not everybody, of course, viewed this closer gearing of the editorial and immensely profitable advertising pages with the enthusiasm of Mr Presbery. 'Time was,' growled Upton Sinclair, 'when you could take the vast bulk of a magazine, and rip off one-fourth from the front, and two-fourths from the back, and in the remaining fourth you had something you could read in the form you could enjoy. But the advertising gentry got onto that practice, and stopped it.'[12]

Indeed they did—and do, although now the exercise is not confined to print. And so the mechanisms and monitors of what we now—with the equivocation which surrounds the subject

—call 'consumer journalism' were institutionalised.

When, a century or so earlier on the other side of the Atlantic, Jeremy Bentham had advanced his moral and political calculus—'the greatest happiness of the greatest number'—he had prudently left his proposition in the abstract. But now in the country where 'the pursuit of happiness' had been enshrined in the Declaration of Independence the requirements were itemised and priced. It would be some time yet before the accountancy of 'Tick which you possess ... washing-machine, refrigerator, deep-freeze, hi-fi, colour TV ...' took full command, but already from the turn of the century the indispensable ingredients of 'the American Standard of Living'* had begun to emerge. It spelt, for instance, 'breakfast cereals'—one of the first great triumphs of the advertising industry—Grape Nuts, Postum Cereal, Shredded Wheat ('Stomach Comfort in Every Shred'), or Force, propagated by 'Sunny Jim' with his 'American Creed', 'I believe that to be happy is all I want,' leading inexorably to the conclusion that 'I know of but one food that makes the creed liveable ... the food that makes me sunny.'†

Equally necessary to American happiness in the swelling cornucopia now spilling from the magazines pages was Coca-Cola, the Victor Talking Machine, a Kodak ('You Press the Button, and We do the Rest'), and the Gillette safety razor, while the much-advertised bicycle was preparing the way for the Pay-as-you-Ride automobile which by 1922 would be the proud possession of one American family in every two.

Enter Social Science with Slide-Rule

The American magazine industry still had much to con-

---

* The Oxford English Dictionary dates the British use of this term 'standard of living,' which was to prove so important a social-statistical artefact, from 1903—in a book by a Scot, A. McNeill: 'The standard of living in England is an ... artificial standard. Practically every Englishman lives, or longs to live, beyond his means.' In America, however, this would already have been accounted a virtue.

† After over two generations in which the absence of this classic item from the American breakfast table would no doubt have defined 'underprivilege', a Government nutritionist, Robert B. Choate, Jr, told the Senate Consumers' Sub-committee in July 1970, that '40 out of 60 name-brand cereals offer empty calories ... they fatten, but do little to prevent malnutrition.'

tribute to the expansion and mechanisation of 'consumer [or producer?] journalism'. In 1911 the Curtis Publishing Company appointed Charles Coolidge Parlin to the novel post of Marketing Director, heading what is generally considered to have been the first market research department in American history. It is a little noticed departure, although a highly significant one for the central democratic issue of true freedom of choice. Parlin not only gathered, analysed, and passed on to advertisers a mass of information about the Curtis publications' readerships, he also travelled many thousands of miles surveying department stores, studying buying—as well as reading—habits. He put in a whole year studying the sales prospects of the automotive industry.

The major publisher, explains Theodore Peterson, in his book *Magazines in the Twentieth Century*, had now 'become a dealer in consumer groups as well as a dealer in editorial matter. He decided upon a group of consumers which advertisers wanted to reach, and he attracted the consumer to the magazine with a carefully planned editorial formula.'

An early and highly successful practitioner, Mr Condé Nast, after spending the first seven years of the twentieth century as Advertising Manager of the mass-sale *Colliers*, launched out on his own with *Vogue, Vanity Fair*, and *House and Garden*. His plan, he explained with commendable lack of hypocrisy, was to 'bait the editorial pages in such a way as to lift out of all the millions of Americans just those hundred thousand cultivated persons who can buy quality goods'.

In establishing the controlling—largely statistical—mechanisms of this silent social engineering, publishers, editors, and advertisers were now reinforced by an advance guard of social scientists. As early as 1900 Professor Harlow Gale wrote *The Psychology of Advertising*, while another academic psychologist, Dr Walter Dill Scott, presided over a Psychological Laboratory at North-Western University where advertisements were tested for 'Attention Value'. In his book of 1908, Dr Scott relates how his 'blindfold test' had established that people could not, in fact, tell pork from turkey, or maple syrup from sorghum molasses. He drew the conclusion that what mattered was the *creation* of the appropriate values—'an atmosphere of elegance and glamour'. He also drew advertisers' attention to the critical importance of habit, pointing out that 'if you bend a piece of paper and crease it, the crease will remain even after the paper is straightened out again.'

In the creation of the appropriate atmosphere, not to mention the creasing and bending, publishers, as well as advertising men, obviously had a role to play. And there could be no room for complacency, no rest from striving. For the higher the structure was piled, the greater the momentum which must be maintained if its total collapse was to be averted. As early as 1935 a Mr William H. Lough, President of Tradeways Inc., was warning Americans that as basic needs were met there was a grave danger that 'the law of diminishing returns of satisfaction' would set in. 'The weakness and limitations of human wants,' he pointed out in his book *High Level Consumption*, 'are completely ignored in many proposals for economic betterment ... Distaste for work, on the other hand, needs no stimulant. It spreads fast ...' Economic disaster impended if 'the great body of consumers' were to be left 'to wait passively for further developments'.

In colonial Africa the problem had been solved briskly enough by a swingeing hut tax; in the democratic America of the Pursuit of Happiness (and 'High-level Consumption') 'desires,' noted Mr Lough, 'are kept alive almost wholly by repeated pungent stimulants; the artifices of dynamic marketing ...'

But what if these, too, should prove subject to some Law of Diminishing Returns? Already, in 1927, we find a marketing research expert asking, a little plaintively: 'What possibilities are there of discovering in him, the consumer, a response that had not yet been appealed to?'[13]

He was not altogether without hope, even though 'planned obsolescence', not to mention credit cards, was still well over the horizon. For social scientists, statisticians with the exploratory drills of 'sampling' technique, now lent their aid. The clipboard lady was about to appear on the doorstep with a thousand questions which would unveil a new universe of desire; in London, Arthur Christiansen, editor of the *Daily Express*, instructed his staff to aim at 'young people with their own car and home, and with a desire to improve their social circumstances, and at young people who do not possess their own car, but have the ambition to do so at the earliest date'.

All ABs Are Equal

The now semi-automated processes of unnatural selection described clearly tended to the 'survival' of the compulsive spenders, and to the social eclipse—failure of adaptation to environment—of any showing a lingering inclination to follow Henry David Thoreau in his belief that he is rich whose wants are few. The image, in short, conformed to the needs of the mirror, and the subject strove to deserve the reflection.

As early as 1914 Dr Daniel Starch of the University of Iowa —the creator of the famous Starch Ratings of magazine advertisement pages—had warned in his *Principles of Advertising* that it was 'obviously useless to advertise in magazines intended for families with small incomes' and had, indeed, included a graph to illustrate the point. Since then the advance of the science of social statistics has made it possible to go beyond the rule of gross numbers of the early Pulitzer–Northcliffe era to readership profiles, including socio-economic class or income-groups, possessions, attitudes, and life-styles. With the aid of the media research men, manufacturers are thus enabled to see which papers are worthy of their support.

Based on scientific sampling, such readership analysis began in Britain in the thirties, multiplying in the fifties when the Institute of Practitioners in Advertising launched its National Readership Survey, designed by the Professor of Statistics at London University, and involving around 16,000 interviews from random samples scientifically taken in 160 parliamentary constituencies.

As the much-headlined percentages of Dr Kinsey, or the deadly decimals of the Growth Rate, or the conclusive verdicts of Box Office were turned into social artefacts by projection in the media, the shape and whole character of the information media themselves was being quietly formed by other— and possibly in some ways equally arbitrary—statistical indicators.

Newspapers flaunted their possession of AB—well-to-do and

middle-class—readers as an actress her minks. '71.5 per cent of our readers have bank accounts, as against the national average of 41.7 per cent,' boasted the London *Sunday Telegraph*, 'You can reach our 1.5 million ABC1 readers for only 1.09 p. per thousand ... Source: Target Group Index ...'

And the lot of those papers rich in ABs could indeed be enviable, for, as Mr John Hobson, specialist in media research, points out in his authoritative work on the subject,[14] an advertiser is 'known by the company he keeps' and 'in their feature articles ... the newspapers afford the same kind of preconditioning of the mood of the reader to advertising as the women's magazines.'

Woe therefore to those improvident papers which, having neglected Dr Starch's early warning, had found themselves with an excess of faithful readers of classes C2, D and E.* To compensate for their readers' lack of per capita spending power the editors of such papers would be obliged to deliver them to advertisers in vast and uniform quantity. If, in addition to its undesirable readers, a paper were to be politically overconcerned, and on the left at that, advertisers might care even less for its company. Thus, the Liberal *News Chronicle* perished in 1960 not so much of too few readers, but of the 'wrong' readers, not properly conditioned for consumption in its pages. And the Labour *Daily Herald*, with only 2·7 per cent of the country's AB readers (as against the *Daily Express*'s 28 per cent) faced a heavy, built-in financial handicap in any attempt to survive as a serious paper of the left. After a long, loss-making struggle and vain attempts to shake off what the adworld called its 'cloth-cap image', it finally succumbed to the attempt to reconcile its purposes and the inflexible rules of the newspaper Numbers Game.

Such statistical monitors, registering both heads and spending power, exerted their steady, silent pressures, with predictable results, building and buttressing a world after the favoured pattern. As early as 1935 in the United States, Gilbert Seldes was noting that 'almost all the newspapers regularly support the presidential candidate whom less than half the people vote

---

* The IPA Survey split 'C' readers into two—C1 'lower middle class' and C2 —'skilled working class.' Below came 'D' merely 'working class' and 'E' the poorest class.

for.'[15] Similarly, with a country almost equally divided between the Conservative and Labour parties, Britain finally had a daily press which seriously represented Labour's philosophy hardly at all, and in which both the ordinary workers' point of view and any consistent radical approach went largely by default. No less serious—under those pressures making for 'quality' consensus journalism—was steady erosion of papers with a true character and distinctive stand, able to enrich and illuminate the national debate by their diversity.

To which, of course, the reply will be made: 'You are both impracticable and out of date.' For have not percipient social scientists, objectively metering the social mirror, long ago announced 'the end of ideology'? Is it not well known that mirrors, like cameras, cannot lie?

'A Latchkey to Every Home'

But this is to get ahead of the story a little—and in particular to omit the catalytic factor which accelerated and deepened these changes, and finally rendered them—or so it seemed— irreversible. This factor—which was powerfully to affect the whole Western world—was the appearance and upsurge of that unique and extraordinary institution, American radio.

Despite the mounting commercial pressures, the intimidating circulation monitors, newspapermen worked in a proud tradition. If the role of the conscientious reporter in relentless pursuit of the Truth has been heavily romanticised, nevertheless the appeal to the sacredness of 'the Facts', to the Freedom of the Press as enshrined in the First Amendment—or echoed in the cry for 'Wilkes and Liberty!'—remained valid enough to ensure that newspapers were often much better than in all the circumstances we had a right to expect them to be.

In spite of the insistent claims of entertainment since the Pulitzer–Northcliffe revolution, journalism managed to retain a basic seriousness of purpose. Radio, by contrast, not only lacked all tradition, but in America was as wholeheartedly dedicated to entertainment as was Hollywood. Secondly, while the Press had become dangerously dependent on advertising revenue, it had retained enough prestige to control and even, on occasion, discipline advertisers. For 'free' radio, however, in the United States, advertising was the sole source of revenue, and despite such Federal watchdogs as the FCC, the logical consequences of this fact were all too obvious. As the president of the NBC observed to investigating members of the Senate's Interstate Commerce Committee in 1943: 'He who controls the pocket book, controls the man.'*

Thirdly—and most significantly for the theme of this book—American radio's commercial rationale lay in its capacity to marshal and deliver mass audiences whose global totals exceeded even those of Hollywood. And this audience was not only vast, it was also continuous in time and space. To monitor its movements was, it seemed, to have a finger on the living pulse of a nation.

Radio was, comparatively, a latecomer. The first regular network was not inaugurated in America until November 1926, and even then extended only as far west as Kansas City. Yet from quite early days it was recognised that, in the United States at least, radio could be a major formative force. It could unite the whole vast land in the same moment of time. And yet despite its universality, its voice was an intimate voice; it offered, as Harry F. David, vice-president of Westinghouse, observed: 'a latchkey to nearly every home'.

'It is inconceivable,' said Herbert Hoover, Secretary of Commerce in 1922, 'that we should allow so great a possibility of service ... to be drowned in advertising chatter.' David Sarnoff, vice-president of RCA, the consortium of major electrical concerns formed to speed radio on its way, considered that broadcasting should be entrusted to a public service organisation, financed by a levy on the sale of sets. He foresaw 'a public institution of great value, in the way that a library is regarded today'.

The first radio conference of 1922 voted that 'direct advertising' on the air should be prohibited and 'indirect' limited to the names of sponsors. Unfortunately, having willed the end it totally failed to provide the means. That same year a 'Gold

---

* And there could have been little doubt about who controlled the pocket book. In 1944 three advertising agencies bought nearly a quarter of the time sold on three of the four networks; half that of CBS in 1945 was accounted for by five agencies and twelve sponsors.

Rush' developed: over 500 radio stations went on the air. Soon there were more than a thousand in clamorous competition for the ear of America. Sales of time rose from a gross $5m in 1927 to $62m by 1933 and $307m ten years later as advertisers reached out eagerly for that 'latchkey to nearly every home'.

As costs escalated in the unremitting struggle to win and hold the mass audiences that alone would justify such charges, the advertising agencies took over the creation of the programmes from the networks, taking their payment, on the newspaper precedent, in the form of 15 per cent of the cost of the airtime—an excellent recipe for further escalation. 'Until 1930,' explained the historian of the N.W. Ayer Agency, 'all agencies tended to look for attractive programmes and then seek advertisers ... After 1930 ... the Ayer firm rapidly developed the view that an agency must start with the client's sales problems ... and then devise a programme which will achieve specific ends in terms of sales.'[16]

According to the historian of American radio, Erik Barnouw, 'by 1932 network approval of agency-built programmes was considered a formality.'[17] One arrived at the extraordinary fact that the radio pabulum of a great nation during all the peak evening hours from 7 p.m. to 11 p.m. was being provided by a handful of advertising agencies, paying out many millions of dollars both to entertainers and to the networks, with the prime object of selling goods.

A major advertiser described his approach thus: 'We give 90 per cent to the commercials and what's said about the product, and 10 per cent to the show.'[18] Entertainers, and even newsreaders, were there to sell as well as perform. And the needs of the sales pitch dictated not only the quality of the programme, but also its form and length. 'Where the listener must be cajoled into immediate action, the programme must be short and snappy.'

Since the bulk of the advertising money came from cheap mass consumer goods, the largest possible audience—to be held against all attempts to steal it—was the prime imperative. The predictable result was a plethora of undemanding entertainment on well-tried formulae. Habit—the creasing of the paper visualised by Dr Walter Dill Scott in 1908—was employed to the point of anaesthesia. By 1950 over a hundred different radio shows had been on the air for a decade or more. The swelling millions of Pulitzer and Hearst, the numerical escalations of Southwood and Beaverbrook, the Box Office astronomy of Laemmle and Zukor, the popular vote totals of Roosevelt and Landon, all were eclipsed by this most brutal, all-pervasive, and ever-present development of the Numbers Game.

There were, it is true, more benign aspects: the 'prestige' programmes, typified by NBC's hiring of Toscanini in 1937, and —in the period unlikely to offer advertisers a large enough audience—the 'sustaining programmes'. But with the advent of soap operas in the thirties, even the daylight hours were largely devoted to gathering, counting and holding an audience for the salesmen. By 1940 the average network was devoting five hours daily to daytime serials succeeding each other in 15-minute instalments, interspersed with commercials. A single advertising agency had no fewer than eighteen soap-opera serials in production simultaneously in its New York plant.

There was now, it is true, a growing band of overseas news-correspondents and analysts. Yet in 1936 H. V. Kaltenborn, reporting the Spanish Civil War (at his own expense), and broadcasting a bombardment from the shelter of a haystack so that Americans might hear the sounds of war, had to wait for a period unsold to any sponsor before he could go on the air. The 'Ratings' ruled, and not until interest quickened in 1938 (the year of Munich) did Kaltenborn achieve sponsorship—from General Mills, the food company. In the same year Raymond Gram Swing received the accolade—from White Owl Cigars. Swing has described his puzzlement at finding that even his friends seemed to think better of him now that he had a commercial sponsor.

In this, however, his analysis perhaps lacked its usual keen-ness. For as Northcliffe, in cryptic, clairvoyant mood, had said of the Daily Mail's new journalism, some twenty-odd years earlier: 'Everything counts, nothing matters.'[19] It was to be the perfect summing-up of the oncoming age of the Numbers Game.

Monitoring the Minutes

In 1929 a statistician named Archibald Crossley devised a method of measuring radio audiences by telephone, and jointly with station operators and advertising agents formed CAB—the Co-operative Analysis of Broadcasting, radio's equivalent of the newspaper world's Audit Bureau of Circulations. Four years later came 'Hooper ratings', the brainchild of C. E. Hooper, a Harvard Master of Business Administration, working with a psychologist partner, Dr Matthew Chappell. Hooper used 'the coincidential' method which meant that instead of telephoning people, like Crossley, and asking them what they *recalled* listening to he asked them what they *were* listening to.

Translated into graph-lines, the Hooper figures not only measured the size of a programme's audience, but continuously monitored it, showing it building up—or fading away—with each five minutes. This way, Hooper claimed, a 'show with a high average rating could be built',[20] by, for instance, cutting out the less high rating bits. A soap opera, shown to be sagging in the middle, could be given a new lease of life with an appropriately inserted suicide or two. The movement of the Hooper ratings was watched with the same mixture of apprehension and awe that in an earlier age would have attended the shooting of comets and similar cosmic events.

All this owed a good deal to the developing links between the worlds of social science and of the mass media. Thus in 1934 a young postgraduate teacher in the psychology department of Ohio State University, completing a doctorate thesis in the field of industrial psychology, posted a copy to CBS. The thesis was entitled 'A Critique of Present Methods and a New Plan for Studying Radio Listener Behaviour'. One conclusion was that advertising was more effective when heard than when merely read. The young psychologist, whose name was Frank Stanton, was promptly hired as a 'listener research' specialist. In 1945 Dr Frank Stanton became CBS's General Manager, and the following year, at the age of thirty-eight, its president.

A pioneer of audience research, Dr Stanton is the co-inventor of the Lazarsfeld–Stanton Programme Analyser, an electronic device for the continuous profiling of an audience's reactions.

Such mechanical monitors were now beginning to impart a new air of precision and inevitability to the 'verdicts of the people'. Around 1935 there appeared a small device, developed at the Massachusetts Institute of Technology, which seemed as timely, and possibly as far-reaching in its possibilities as that earlier American invention, the Hollerith punched card tabulator machine. The Audimeter crouched beside the family radio and silently recorded on tape the listening behaviour—the verdicts —of the set's owners. Audience gains and losses could now be shown, and totalled, minute by minute.

By 1943 the Nielson market research organisation had installed its own version of the Audimeter in 800 homes in the North and North-Central regions of the country and was claiming to offer a radio audience profile representing one-quarter of the radio homes of the United States—a moment-to-moment summation, surely unique in history, of the continuous, collective 'verdict of the people'.

In the preceding chapter we have seen how the statistical revelation of Thomas Malthus, Charles Booth, Alfred C. Kinsey and others 'made history' in both the literal and metaphoric senses of the phrase. Now new statistical artefacts were appearing in the cultural, as in the economic and political field, to shape and contain our lives. But unlike the earlier statistical 'moments of truth' they were often *continuous* 'indicators'. And they were arrived at by statistical and mechanical processes too detailed and protracted and too boring to make visible.

Their authority was not diminished. On the contrary, their numerical verdicts appeared to reflect the same sort of inevitability as the statistics of Malthus. As Stanton pointed out: 'A mass medium can only achieve its great audience by practising cultural democracy—by giving the majority of the public what they want.' And what this was the totals of the audimeters now crisply indicated and the statisticians analysed with all the scientific assurance of surveyors making a triangulation of a series of peaks.

In fact, of course, the statistical procedure and approach were all the time enlarging and reinforcing the 'mass' in 'mass medium'. The process was at least in part circular.

But there seemed no way out of the circle. Certainly, the coming of television didn't offer one. In Britain it was television which 'went commercial', an event which brought in its train many of the processes of automated pseudo-democracy pioneered in America, and breached the defences of the old élitist society. In the United States television, like Hollywood before it, stepped up the showbiz fantasy, further inflated costs and heightened the salespitch. If the networks again provided the programmes, it was the sponsors who determined whether they should go on the air. The tyranny of the ratings was intensified: all depended on it. The price of the 'Dr Kildare' series rose by over 50 per cent as its ratings rose.[21]

Since the sums at stake were vast, anything which could put the ratings at risk was clearly objectionable. It can come as no surprise that in 1950 during a recrudescence of the Great Red Scare scores of talented writers and entertainers were driven off the air on suspicion of possessing left-wing sympathies.* 'Controversial people,' it was explained, were 'bad for business.'

On that occasion Raymond Gram Swing warned of the danger of such 'pressure-groups of blacklisters' producing a 'single conformity of thinking in America.' Yet he remained confident. And the way in which he expressed this confidence is interesting: '... democracy is a free market ... democracy is, among other things, a belief that in this free market the unworthy will ruin themselves, and the truth in time—yes, only in time—will triumph.'[22]

Dr Gallup's 'Continuous Audit'

'Free market' or not, democracy in America was certainly developing accounting instruments of much versatility and central importance which were far indeed from the dreams of the Founding Fathers, and still outside the scope of most textbooks of political science.

In the year which marked the 'audimeter's' experimental debut as a cultural arbiter, Dr George Gallup founded his American Institute of Public Opinion, and on 20 October 1935, published the first Gallup Poll in history. Dr Gallup himself has

---

* The whole extraordinary story is detailed in volume two of Erik Barnouw's US radio history, *The Golden Web*, Chapter 5: 'The Purge'.

described his opinion poll as 'the marriage of journalism and sociology'. But a reading of this chapter may suggest that it was, in reality, more in the nature of a *ménage à trois*. The third partner, discreet, but not impassive, was marketing and advertising.

Here again, now extending into the political field, was one more product of that quiet, but developing working alliance between journalism, merchandising, and social science which since the turn of the century had been perfecting the mirror in which society saw its image. Most of the pioneers of this time blend these same elements in their careers. With a PhD in psychology from the University of Iowa, George Gallup had worked on the *Daily Iowan*, and had taught journalism at three universities. But he came to opinion polling from a research post at the advertising agency, Young and Rubicam. Gallup's rival pollsters, Elmo Roper, who started the *Fortune* survey in 1935, and Archibald Crossley of 'Crossley Ratings' fame, were also from advertising.

But if journalism and market research each made a large contribution to the evolution of the opinion poll, the element which ensured its brilliant future derived from its social science side—more particularly, from its technical base in the science of social statistics. This was unforgettably demonstrated by Dr Gallup himself at the outset in 1936, when he not only flatly contradicted the predictions of the *Literary Digest*'s celebrated straw-vote—an American institution since the First World War—on the Landon-Roosevelt Presidential election, but also foretold the *Digest*'s error to within one per cent.

The *Digest*'s prediction was based on ten million mock-ballot cards mailed to its subscribers and to homes on the telephone. Dr Gallup's prediction was based on a scientifically constructed 'stratified random sample' of the electorate, totalling only a few thousand. Yet the *Digest* got it disastrously wrong, and Gallup got it right*—a landslide to Roosevelt.

Rarely since the vast life insurance industry rose on the foundation of the Expectation of Life Tables can the science

* He was also lucky. His actual error in the prediction of the party popular vote was 6.5 per cent. In 1948 a smaller error than this—i.e. an error of 5.3 per cent—resulted in the notorious false prediction of the defeat of Truman.

of statistics have received a more impressive vindication. It was soon evident that here was a social invention—a new statistical artefact—of far-reaching possibilities. The most fervent of democrats, in the tradition of the Levellers or the Chartists, had merely called for annual elections. But now Dr George Gallup and his colleagues were offering to provide what Dr Gallup was soon calling 'a continuous audit' of public opinion.

The phrase 'audit' no doubt came naturally to one whose faith in 'democracy as a free market' seemed to be as robust as that expressed by Raymond Gram Swing. 'Shall the common people be free to express their basic needs and purposes?' Dr Gallup demanded in his book, *The Pulse of Democracy*, in 1940, 'or shall they be dominated by a small ruling clique? ... Democracy recognises the essential dignity of the individual as such; it assumes that our economic, political and cultural institutions must be geared to the fundamental right of every person to give free expression to the worth that is in him.'

So the opinion polls would provide democracy's gearing. But gears imply driving shafts and motive power—and it was becoming less and less clear, as the 'gears' multiplied, where this might lie.

On this matter, however, events in Britain in the fifties and sixties were to cast a certain amount of light.

4

POP GOES BRITAIN

Democracy to complete itself; to go the full length of its course towards the Bottomless, or into it ... Count of Heads to be the Divine Court of Appeal on every question and interest of mankind ...
> Thomas Carlyle, *Shooting Niagara: And After*, 1867.

A programme in which a large part of the audience is interested is by that very fact ... in the public interest.
> Dr Frank Stanton, President CBS, to the Federal Communications Commission in 1960.

... the fourth great revolution in public communications.
> Sir Robert Fraser, first Director-General of the Independent Television Authority, on the introduction of Commercial Television in Britain, 1955.

The Gallup Poll had arrived in Britain in 1936, and the following year had been taken up by the Liberal *News Chronicle*. But no one paid much attention to it—not even when, seven years later, it foretold the unthinkable, a postwar Labour victory at the polls. For Britain was still clearly a representative, rather than a mass—or mass-market—democracy. Thomas Carlyle's grim forebodings at the extension of the franchise in 1867 had remained largely unfulfilled; Niagara was still unshot in a country where, on the whole, opinion was still made by 'the conversation of the people who counted'.

It is true there were growing signs that beneath the surface of traditionalism, the 'Divine Court' of 'Count of Heads' might be gaining ground. Although in theory despised, Hollywood's Box Office values continually made the pace and the headlines. In the great circulation war of the thirties the *Daily Express*, of transatlantic inspiration, had emerged a clear victor, although in championing the 'Little Man' and the freedom of the suburban consumer and scorning effete 'lords', it was somewhat hampered by the aristocracy's high gossip-column value, not to mention Lord Beaverbrook's own barony. Also advancing vigorously in the circulation stakes, from the other side of the social spectrum, the 'new, radical' *Daily Mirror* cocked a tabloid snook at what, in the year of Munich, it had called 'the upper-class riff-raff', but by the very eagerness with which it called a spade a 'bloody shovel', showed that it too knew its place.

So did most Englishmen, and, despite ritual Left Book Club protest, appeared content enough to do so. The social sciences, which, in America, were developing the feed-back monitors of control, and lending the prestige of science to marketing, were in Britain still highly suspect, their very existence barely acknowledged at Oxford. Westminster still rejoiced in the title of 'Mother of Parliaments', the calm of that tolerably good club, the House of Commons, remained relatively little disturbed, and even the 'German Dictator' seemed much less alarming when encountered after the strokes of Big Ben in the BBC's Nine O'Clock News, read by dinner-jacketed announcers whose every vowel and inflexion had been predetermined by a committee of 'the best and most experienced minds'.

George Bernard Shaw, who was a member of the committee, had, it is true, once suggested to Sir John Reith, the BBC's Director-General, that it would be improved by an age limit of thirty and the addition of a few taxi-drivers.[1] Yet if the spoken English of its announcers bore little resemblance to the sounds on the lips of the greater part of their fellow citizens, nevertheless, in some curious way, the British Broadcasting Corporation emerged as a truly national institution in which even its most persistent critics were eager to assert proprietory rights, while millions of the more humble gave it the trust and familiar affection they might have given to a favourite aunt. The centre held.

The parallel with the evolution of radio in America is instructive. For, as in the United States, the starting point in Britain also had been a consortium of radio manufacturers, principally concerned to promote the sale of their wares. Yet in Britain the public service organisation which RCA's David Sarnoff had visualised for America had actually come to pass. The reasons for this dramatic divergence from the common starting point are, of course, multiple. They include the vastly different geographical characteristics and social structures of the two countries, and the strength of the hierarchic and public service traditions in Britain as against the hardly less strong idea of 'democracy as a free market' in America. But the decisive factor was probably accidental: it lay in the character and sense of mission of one man, John Reith, the 34-year-old Scots engineer who in the year of Lord Northcliffe's death, applied for the post of General Manager of the newly formed British Broadcasting Company.

Almost single-handedly at first, by the strength of his faith in broadcasting, and by sheer force of will, John Reith (later Sir John, then Lord Reith) transmuted this commercial venture with strictly limited aims into—what was then a novel type of social institution—an independent public service corporation with wide horizons, duly regularised by the grant of a Royal Charter in 1926. For years after that, Reith fought off or skilfully deflected threats to the BBC's integrity from leading politicians, from big business, from the counters of heads and the counters of money, until the organisation—or the 'organism' as he, significantly, preferred to think of it—had taken on a life of its own.

Could broadcasting, Reith asked, become 'the tempering factor that would give democracy for the first time under modern conditions a real chance of operating as a living force?' Certainly, to use radio for light entertainment alone would be 'a prostitution of its powers and an insult to the character and intelligence of the people'.[2]

Reith never felt the slightest need to apologise for what he was doing. 'It was,' he wrote in his autobiography, 'the combination of public service motive, sense of moral obligation, assured finance [from radio licences] and the brute force of monopoly which enabled the BBC to make of broadcasting what no other country in the world has made of it ...' Particularly 'the brute force of monopoly'—without which, he explained, the BBC 'might have had to play for safety, prosecute the obvious popular lines ... count its clients ... curry favour ... subordinate itself to the vote.'[3]

Reith, in short, was almost as determined an enemy of 'Count-Heads' as his fellow-countryman, Carlyle. To those who accused him of being anti-democratic, of giving the public, not what it wanted, but what the BBC thought it needed, he replied: 'Few know what they want, and very few know what they need ... in any case, it is better to over-estimate the mentality of the public than to under-estimate it.'[3] And it can indeed reasonably be argued that Reith was a truer democrat than many of his accusers—because it was basic to his policy that, not merely an élite, but all men could, should—and indeed would—enjoy the best, given time and opportunity to do so. If he was perhaps too much a realist for the Tennysonian 'we needs must love the highest when we see it', he nevertheless stuck firmly to his early declaration that 'the policy of the company is to bring the best of everything to the greatest number of homes.'

But if the number of homes switched on does not define the best, the prime difficulty of such a policy is to know how then to identify it. In an élitist society, it is apt to boil down to 'the classics' or the consensus of received opinion. Hence the notorious identification of the prewar BBC with the interminable 'Foundations of Music' series which, together with Reith's equally notorious sabbatarianism, celebrated in innumerable Bach church cantatas, admittedly drove substantial numbers of British listeners across the Channel to the commercial stations of Luxembourg and Normandy.

Yet for most of the time Reith's high claims for the BBC were justified. From early days the BBC's modestly paid but dedicated staff provided a balanced and wide-ranging service with something worthwhile for most tastes. And they did this, it might now be noted, without any systematic provision at all for checking on public reaction, much less monitoring and charting 'what-the-public-wants'. The Reithian BBC simply challenged its professionals to be professional—to set their own high standards.

However, the need to know more about audience reaction than listeners' random letters and remarks revealed was obvious enough, and in 1936 a new Public Relations Controller at the BBC imported a young social statistician from the market research section of a London advertising agency and gave him the job of organising an Audience Research department. But, significantly, Robert J. Silvey's various 'barometers' and 'thermometers' of audience reaction did not become fully operational until December 1939—more than a year after Reith had left the BBC.

From his new situation Reith continued to issue sombre warnings against these head-counting devices. 'Listener Research,' he told the Beveridge Committee on Broadcasting, in 1949, is 'subversive, and a menace as here practised. On the scale and system now in operation' it was a 'waste of time and money; its results are unreliable and misleading; it is inevitably a dragdown; it causes producers to look to itself [i.e. instead of to their own professional consciences] for the criteria of success.'

The Masterful Servant

But if audience research was 'subversive', it was by no means the only subversion at work in the England of the forties. Indeed Reith himself was certainly fortunate in the timing of his departure from the BBC—just when a 'dictator' even more single-minded than he was about to take over the life of the country. The war brought full-fledged 'popular radio' to Britain. The first British soap opera, 'Front Line Family', appeared in 1941 and with one change of name, soldiered on, six times weekly, until 1947. 'Tap-listening' had formerly been frowned on; the need to combat the boredom of blackout or fire-watching made it routine. 'Music While You Work', 'Workers' Playtime', and similar undemanding fare jingled interminably from canteen loudspeakers. American comedy shows—'deloused', or divested of their commercials but not of their canned laughter—now became for the first time familiar to the British listener. Their pace and professionalism made its impact.

The detailed statistics of popular reaction which had seemed so sapping to Reith were now, in the conditions of wartime, beginning to appear the key to effective administration, to put it no higher. In 1941 the British Government set up a permanent Social Survey to brief departments about the impact of their activities on the people. The BBC's 'Music While You Work' programme had, in fact, been tailored to the reports of war factory managers. In one way or another, whether by

setting up social 'feed-back' devices, or through the investigation of morale or productivity, or by the use of psychological
testing in the selection of pilots and others, the social sciences
vastly enhanced their status in Britain in these years.

At the BBC, Audience Research could at last come into its
own. Its distinctive feature, its designer, Robert Silvey, explained, was that it applied 'the scientific method'. Employing
'aided recall', the technique originally used by Crossley in
America, the daily 'Listening Barometer' was constructed by
interviewing a scientifically chosen representative sample of
400 listeners in each of the BBC's seven regions, an unending
process involving one million interviews a year.[4]

By 1942 the numerical verdict of Silvey's meters was crisp
and inescapable. They showed that the Forces Programme, with
its large Americanised entertainment content, had drawn a
civilian audience half as great again as that for the Home Service
programme. The British people had, in a sense, 'voted' in a
way they had never been invited to vote before. When peace
came it was clearly no longer possible, in a democracy, to revert
to the full Reithian rigour.

BBC NOW A YANKEE BABY IN CARBON COPY OF US PROGRAM-
MING, ran an exuberant headline in *Variety* in 1948. It was
premature, but not without point. The Forces Programme had
been civilianised into the Light Programme—and in Reithian
terms that very category spelled abject surrender. But there
was also the more substantial Home Service, and, further to
restore the balance, there was added a year later a Third
Programme, designed to offer the best in the arts and sciences
to those—few, it was now rather implied—capable of appreciating them.

It was an idea Reith had himself already considered and
rejected in the thirties, when he had witheringly condemned
such minority 'Culture' programmes as 'an easy way out'. Even
so, the postwar compromise by no means wholly discarded the
Reithian ideal. The three alternative programmes were described by the author of the New Model, Sir William Haley, a
former newspaper editor, as 'a broadly based cultural pyramid,
slowly aspiring upwards ... The listener must be led from the
good to the better by curiosity, liking, and growth of understanding.'[5] Thus, in theory at least, the BBC, this central, and

somehow characteristic, institution of contemporary Britain, still embodied 'representative' rather than 'mass' democracy.

Yet Reith's sensing of the 'subversive' power of a continuous head-count had been correct. There was now the constant presence of what one old-line BBC man called 'the Abominable Statistic'.[6] The findings of audience research, it was repeatedly stressed, were never mandatory. Indeed, the numerical totals, the scrupulously calculated percentages, the crisp indices were kept secret from the outer world. They were to be 'the servants, not the masters, of programme planning'.

All the same, as Maurice Gorham, Head of the BBC's Light Programme wrote later: 'it was the sort of servant whom nobody could afford to displease.'[7] He wrote with feeling, for, in fulfilment of the Haley 'pyramid' design, he had been given the task of attracting, for the Light Programme (the pyramid's base) 60 per cent of the total British audience. It was the beginning of the BBC's Numbers Game. Gorham describes vividly in his book how the rival outfits of programme-planners 'pored anxiously over the daily listening figures', like chess-players anticipating the opponent's moves, and fearing imminent checkmate. Gorham's particular preoccupation at the Light Programme was to find the item which would somehow 'beat' the Nine O'Clock News on the Home Service.

Even so, the BBC did not allow itself to be bound to the wheel of undiluted 'count-heads' without a struggle. Alongside the index of audience numbers was set up a qualitative, or appreciation, index, elaborately calculated from the personal judgments on programmes by representative listener panels in each of the BBC's seven regions. Each of the total of 4,200 panel members marked each item with his own degree of liking or disliking on a five-grade scale, running from 'A+' for 'quite exceptional' enjoyment through 'B' for 'normal' enjoyment to 'C−' for 'extreme distaste'. With 'A+' counting as four points and 'C−' as zero, an index was calculated showing 'the average mean reaction' which was expressed as a percentage of the maximum possible appreciation (if everybody had voted 'A+').

The final index figure representing, so to speak, the relative 'net enjoyment' of each programme, was, of course, admirably precise (and might even be accompanied by some verbal comments). But, as with so many such heroic efforts to save the qualitative by armouring it in the quantitative, it tended to be self-defeating. It lacked conviction. For could the 'quite exceptional' appreciation experienced by one person really be considered equivalent to the 'normal' appreciation of two persons (2+2 pts=4 pts for a single A+) or to be cancelled out by the 'extreme distaste' experienced by, say, Colonel Blimp? What if the Colonel positively *enjoys* experiencing 'extreme distaste?' 'Dislike,' commented the BBC's Audience Research director, Robert Silvey, 'is often deepened by listeners who "have got into" the audience for the programme by mistake, and are, in fact, the "wrong audience",[4] but it seems, lack the energy to switch off.'*

Not surprisingly, programme planners tended to prefer the Abominable Statistic in its less refined forms,† the grand total of switched-on heads. And crude or not, it was a measure that had an intimidating clarity in a land formally dedicated to the political principle of 'one man, one vote'. By 1959 it was showing that the Third Programme's percentage of the audience had dropped from 4 per cent (1946) to 1 per cent in 1949 and then slumped to a figure (36,000) too small to be reliably assessed by the sampling techniques available to Audience Research.

Some, it seemed clear, were failing to love the highest when they saw it. Indeed, if the figures for the Light Programme were to be trusted—and who could distrust figures so elaborately, so scientifically, garnered?—the BBC's cultural pyramid was aspiring downwards.

'Efficiency' in Excelcis

The emergence of such numerical indices of public taste—

---

* Silvey further admits that, in their amiability, listeners use the 'A+' grade much more often than they ought, and feels 'little doubt that many panel members use "A" when strictly they should use "B",' but claims that they will err the same way from day to day, so comparisons of programmes are unaffected.

† Cf. Stuart Hood, one time Head of BBC Television, who refers to the Appreciation Index as 'of little practical value except for a director or producer to clutch at when his programme has been a flop in terms of audience or professional achievement'. (Stuart Hood, *A Survey of Television*, 1967.)

the ballot of the radio switch—was accelerated by the spread of television across the land, leaving a trail of broken habits in its wake. By comparison with radio, television production was complicated and costly. It lived on movement and was impatient of words. It had an insatiable appetite for new faces, 'gimmicks', visual novelty of any kind. Uncouth sounds, with faces attached, began to break through the once well-modulated tones of the BBC. In a somewhat uncertain way, the thing was an equaliser, like the Colt revolver.

In such a dissolving world numerical indices provided, or so it seemed, hard 'reality' to cling to. Particularly since in the early postwar years numerical economic 'targets', percentage increase in productivity, growth rates, the 'dollar gap' and so on had infiltrated deeply into the English language. Repeated everywhere on huge posters, Prime Minister Attlee's clarion call, TEN PER CENT MORE WILL TURN THE TIDE, was, in the face of 'Britain's economic Dunkirk', the contemporary equivalent of Henry V's appeal to British patriotism on the field of Agincourt.

With the return of a Conservative Government in 1951, and escape, at last, from the grip of wartime austerity, the television set became the centrepiece of a whole galaxy of 'consumer durables' and 'emulation goods', fuelling a light industries boom. At the same time the régime of economic indicators and 'targets' was spectacularly supplemented by the excitements of a new financial Numbers Game, played out on the national stage. From the early fifties onwards, the names of the Napoleons of 'take-over' were rarely out of the headlines, as they ran up their empires, generating millions by their operations—in order to take over yet more vulnerable concerns and generate yet further millions. In 1953 Mr Harold Macmillan, Conservative Minister of Housing and Local Government, dropped the property 'development charge' which the Labour Government had imposed in hope of recovering for the community some part of the vast increment in site values its activities brought about. Macmillan explained: 'The people the Government must help are those who do things: the developers who create wealth whether they are humble or exalted ...'

And indeed in the next decade or so property developers were to 'create wealth' on such a scale that in 1967 Oliver Marriott, editor of the London *Times Business News*, was able to print a list of 108 men and two women 'each of whom on my calculation must have made at least £1m' by property development in the 'golden period' between 1945 and 1965. 'Few of these,' added Marriott, 'started with any capital beyond the odd hundred pounds ... all the money to buy the site was usually lent by the banks, all the cost of construction carried by the contractor or the bank, and the total repaid from a long-term mortgage borrowed from an insurance company.'[8]

'Property', which had once supplied at least one sort of basic and abiding value, now increasingly became an affair of manipulated counters, flashing rows of noughts hardly real enough to merit denunciation by Proudhon as 'theft'. The prevailing hero of the headlines was no longer the engineer, the scientist, nor even the *entrepreneur*, but the *prestidigitateur*, the manipulator of 'assets', the runner-up—one could scarcely use the word 'architect'—of the towering 'conglomerate' or financial 'trust'—the 'creator of wealth' in the banker's sense, who could make, say, 9.52 pounds or dollars grow where only one had grown before.

Any surviving inclination to question the inevitability or the social value of such masterful operations, or to inquire whether the vast and often precipitate disruption of lives and organisations involved* was really balanced by consequent social gains, was now effectively nipped in the bud by instant appeal to a few 'accepted' statistical measures of 'Efficiency'— yield on capital employed, yield per square yard of shop or factory floor, rate of earnings growth ...

This fast-expanding world of 'Everything counts, nothing matters' was without natural frontiers. Cultural values too were inevitably subject to the same numerical tests. For 'Efficiency' is a wonderful argument-stopper. Useless, for instance, to point to the toll of human frustration built up by highly efficient assembly lines, about, say, the long delays at the highly efficient supermarket check-out gates or, even, for that matter, about the way in which equally reputable firms of accountants

* As this sort of financial 'rationalisation' increases, firms may repeatedly suffer such disruptions. Thus, the concerns rapidly brought together in the holding company, Fordham Investments, by the much acclaimed '25-year-old whizz-kid', David Rowland, were later taken apart again and redistributed when he sold out his own holdings for £2.4m and the holding company was absorbed by yet another holding company, Ralli International.

can arrive at very divergent figures for a firm's profits—thus presumably raising or lowering its 'efficiency'.*

Thus, by the mid-fifties in Britain, one of the first principles of the Numbers Game was well on the way to total acceptance: that a verbal abstraction is merely an abstraction but a statistical abstraction is objective reality itself. Soon, when in their newspapers the British read that in the company 'growth rating' league table (reprinted from the magazine *Management Today*), the financially-orientated holding company, Slater Walker Securities was 'topping the top ten with a growth rating of 4563 per cent' (calculated on accumulated dividends plus gain in share price over a decade) it would seem superfluous to look further or to inquire what was actually *produced*. Mr Jim Slater, much-sung hero of the financial pages, was clearly a paragon of Efficiency.

The Other Battle of Britain

Of the BBC's pioneers its historian, Asa Briggs, has written: 'They did not hesitate to oppose tendencies which are now thought to be "inevitable" tendencies of our age ...'[9] It is, however, characteristic of the Numbers Game that such independent choices are slowly but surely eroded; the 'inevitable' becomes ... inevitable. Sapped by the 'Abominable Statistic' from within, the BBC citadel was now to fall to the sustained assault of the yet more insidious statistics without.

From the early days of the BBC onwards, a succession of public inquiries had considered, and firmly rejected, the idea of

* This is, of course, rarely publicly demonstrated since rival firms of accountants do not normally operate on the same firm's accounts. But in the dispute over the Pergamon Press, the difference between the true profits figure of two famous firms of accountants was £1,600,000 on a maximum of £2,100,000. According to the Wall Street expert, 'Adam Smith' (*The Money Game*, 1968) a company's net profit can 'vary by 100 per cent depending on which bunch of accountants you call in ... and all this without any kind of skulduggery you could get sent to gaol for.' He instances, amongst other things, wide differences in depreciation accounting. In November, 1971, reporting on 'conglomerates', the US House of Representatives Anti-Trust Sub-committee, chaired by Congressman Emmanuel Celler, wrote of Litton Industries' carefully cultivated image of technological and managerial supremacy '... Litton has utilised all of the sophisticated accounting techniques and statistical gimmicks available. It is adept at concealment, misdirection and incomplete statement.'

permitting advertising on the air as likely 'to lower the standards' (the words of the Sykes Committee of 1923). In 1951 the latest of these inquiries, the Beveridge Committee on Broadcasting, after sifting a vast amount of evidence, had concluded (with four dissentients) that 'if people of any country want broadcasting for its own sake, they must be prepared to pay for it ... they must not ask for it for nothing as an accompaniment of advertising some other commodity.' And with only one dissentient they had likewise ruled that this 'most pervasive and therefore one of the most powerful agents for influencing men's thoughts and actions' should remain in the trusted and highly experienced hands of the BBC.

However, in the autumn of 1951 a Conservative Government was returned to Westminster after fighting an election under the slogan, 'Set the People Free'. In May of the following year a White Paper announced that 'in the expanding field of television provision would be made to permit some element of competition.'

The language was designedly obscure, since the area was felt to be a highly sensitive one. Yet the White Paper was, in fact, to herald one of the most passionate and crucial public debates in Britain's postwar history. It was a battle fought out in that critical, but little-charted, area where the political, economic, and cultural constituents of a society meet and obscurely mesh. Both sides fought under the banner of democracy, yet differed fundamentally on what the word implied. Both fought in the name of freedom, but were very far from agreeing on what they most wished to be free from. It was a division which cut deep across party lines; there was a half-caught sense of a clash of values and systems never adequately defined—of a fated, but imperfectly understood, parting of the ways.

For the traditional ruling party the issue was to prove catalytic. Almost all the Conservative Party's old leadership was instinctively against the idea of advertising-financed television in Britain. Lord Halifax, former Viceroy of India, Foreign Secretary and leader of the House of Lords, declared it 'profoundly wrong'; Lord Brand, speaking from his American experience, complained that sponsored television 'spreads the idea that everything is for sale'; Lord Radcliffe, a Law Lord, warned that the 'level of the culture of the country' was being 'put in hazard'; Lord Hailsham—by no means wholly in jest—compared the proposal to introduce commercial television to the setting up of a Golden Calf.

From the Labour side, the veteran party leader Herbert Morrison declared his deep conviction that 'the projected development is totally against the British temperament, the British Way of Life, the best—even reasonably good—British traditions'. Certainly, in installing the Profit Motive in this critical area it ran full in the face of the party's philosophy, and Mr Attlee had no hesitation in pledging repeal on Labour's return to power.

How then did this revolution—or counter-revolution—for it seems that no lesser term will do—come about?

Revolutions are usually made by small groups of men distinguished by knowing just what they wish to accomplish. This revolution was no exception. What was unusual perhaps was the character of the group—from the advertising industry, from the booming world of public relations, from the radio manufacturers, and a few mass consumer-goods concerns who sensed television's vast potential as a marketing medium at this juncture in Britain's history. They had a few spokesmen in Parliament, and one valuable—perhaps crucial—ally in the Cabinet.

Frederick Marquis, Lord Woolton, the Conservative Party's organiser of victory after the debacle of 1945, did not fit any of the old stereos of the Tory chieftain. A scholarship boy at Manchester Grammar School, he had gone on from there not to Oxford or Cambridge, but to Manchester University, where he had read chemistry, physics, mathematics—and psychology. He cherished the idea of becoming a sociologist, and had actually joined the Fabian Society. For a time he worked as a journalist, then, much in the tradition of Charles Booth, and in Booth's own city, Liverpool—of which Woolton is a suburb—had joined a dockland settlement with a plan to investigate the causes of poverty. There he was awarded a research fellowship in economics, winning an MA for his thesis, and a Fellowship of the Royal Statistical Society.

In the First World War he was recruited as an economist to the War Ministry, and while working on the control of raw materials, came into close contact with business and businessmen. He now concluded that what was required was—in the American phrase—not more government in business, but more business in government. He resigned from the Fabian Society. But the decisive turning point seems to have arrived in 1920 when he accompanied on a business trip to America a member of one of the two Jewish families who owned Lewis's, a Liverpool departmental store.

In his autobiography, Woolton describes in rapturous tones how he saw 'the great store catalogue of Montgomery Ward and how, 'after walking through what seemed miles of stores ... I saw the tremendous effect that retailing might have on the standard of living of a locality.'[10] The young science graduate who had set out to discover the causes of persisting poverty had found a new mission. He returned home, wrote a report on American retailing, and joined Lewis's Stores. This mission at least may be accounted wholly successful. By 1939 when Sir Frederick Marquis, then chairman and managing director of Lewis's, was elevated to the peerage, the storage group had assets of over £5 million. In 1942 the former young Fabian of a generation earlier was brought into politics to fill the post of Minister of Food in the wartime cabinet; in 1942 he became a Companion of Honour, in 1953 a Viscount, in 1955, an Earl.

Despite the medieval trappings, Woolton represented something very new in the centres of British political power. He echoed in his person the bland amalgam of social science and salesmanship and mass-communications expertise already noted as inspiring the design of the statistical instruments of consumer 'democracy' in the United States. He put the Conservative Party into large-scale, slick advertising campaigns. He was himself the prototype of the 'new men' brought into the party by his modernising reforms as its chairman. Like him, they came from the professions and business, journalism, advertising and public relations. They too believed in the mystique of professional management, the power of 'merchandising', the social sciences, market research, efficiency, growth, and the profit dynamic. If in the commercial television debate 'Americanisation' represented a spectre for the old-line Tories, for the new men it was a beckoning goal.

And, in a sense, time was on their side; for the Numbers Game, as we shall see, is both subtly and crudely self-validating. 'Need we be ashamed of moral values, or intellectual and ethical objectives?' asked Reith in the House of Lords. Certainly it was a question which might well earlier have come from the lips of Lord Woolton, at the outset a striking case of the Nonconformist conscience at work. But now the statistics of 'high level consumption', of 'consumer democracy', made their own answer. Before their continually reiterated authority merely individual human judgments began to appear an impertinence.

By cleverly labelling their proposal '*Independent* Television' the commercial lobby classically masked the whole issue of the ultimate, effective control, and kept the critical spotlight firmly focused on the BBC's monopoly.* A former BBC Controller, Mr Norman Collins, now discovered commercial television to belong to 'those Four Freedoms which less than ten years ago we were saying we were prepared to defend with our lives'; while Sir David Eccles, soon to be Minister of Education, assured his audience at a Conservative fête that Stalin, Hitler, Mussolini and Franco would have been against an 'alternative' television programme. In any case, Sir David added: 'Our firms show great restraint and good taste. Our advertising would not be vulgar.'[11]

Foreclosure

So on 22 September 1955—a year which in terms of alien conquest may yet rank with 1066—the Golden Calf was duly set up, 'alongside the Established Church', in Lord Hailsham's phrase, and the wondering populace danced around it, awaiting the magic moment of the first TV commercial in Britain's island story.

The gilt, certainly, had from the first the lack-lustre look

---

* In fact, of course, the issue was not—as so often presented—that of an alternative independent service, in addition to the BBC's—but of the introduction of advertising-financed television. It was incorrect to suggest that an alternative could only be financed by advertising, since the Post Office even then was still retaining 15 per cent of the licence fee. But other possibilities were in effect excluded from debate by the adroitness of the advertising lobby. The issue—crucial, in a democracy—of the finance of mass communication has remained largely unexplored to this day.

of the 'British genius for compromise'—or infinite capacity for self-deception. Not only was 'sponsoring' in the American style forbidden, but so was all participation of advertising agencies in the programme companies. These, it is true, would sell time to advertisers much as newspapers sell space, but the 1954 Television Act ruled that the amount must not be 'so great as to detract from the value of the programme as a medium of entertainment, instruction and information'. Furthermore, the supervisory public authority that was set up, the Independent Television Authority, unlike the FCC in America, was responsible for providing the stations and transmission facilities, besides leasing time on these to the programme companies. Theoretically at least, it thus possessed ample power to carry out effectively its statutory duties of supervising both advertisements and programmes, the first for honesty and good taste, the second for 'balance of subject matter', impartiality in controversial matters, and maintenance of a 'proper proportion' of British material.

The British, explained the Postmaster-General, Earl de la Warr, 'would not respond favourably to the same vulgarities and horrors, and even tiresomeness, that are apparently so popular elsewhere'. 'Needless to say, at no time would interruption of programme by advertising matter be permitted.' [This refers to the higher metaphysical doctrine of 'natural breaks' naturally occurring at approximately 15-minute intervals.]

Across the Atlantic, the trade paper, *Broadcasting*, recording these elaborate arrangements for the segregation of the salesman, was inspired to comment in verse:

> Dear little John Bulls,
> Don't you cry;
> You'll be full commercial
> Bye and bye.

For the Americans understood the mechanism and the relentless logic of this branch of the media Numbers Game. They noted that the British programme companies were normal joint stock enterprises, dependent for their profits, dividends, share prices on the sale of advertising time, mainly for cheap mass-consumption goods, and that this, in turn, depended on the guaranteed delivery of the largest possible captive audience—

commercial television's unique product.

And they knew that however much the Television Act might strive to insulate the programme planners from the salesmen, America had now given the world a fully proved instrument for gearing them together again. Monitoring sets throughout Britain, the 'Tam'-meters (Television Audience Measurement Ltd), unsleepingly gathered the data for a minute-by-minute ITV/BBC audience profile each week, accompanying it with figures showing the cost per 1,000 homes for each commercial.* Through its wired TV-service subsidiary, one company, Associated Rediffusion, was indeed able to go one better with a device called IBAC—Instantaneous Broadcast Audience Counting—which showed immediately the switches from one channel to another, summing them up in a curve as they occurred.

Whereas the BBC, in pursuit of its principle that audience research should be the servant, and not the master, of its creative activities, had not published its 'head-count', the commercial companies inevitably blazoned abroad ratings which were almost as good as cash in the till. They revealed a graph-line rising like Table Mountain for giveaway quizz shows like 'Double Your Money', or for hospital melodrama serials, or for 'Wagon Train', but plunging to Dead Sea levels for, say, the Halle Orchestra or Shakespeare. So now, if programmes were in some degree 'balanced' by some more significant—or 'prestige'—material, this was placed late or early, outside the hours when most people could watch and offer that relaxed, off-guard attention in the privacy of their own homes for which advertisers were willing handsomely to pay.

In short, this small monitoring device, the gift of American ingenuity and 'know-how', secured on the one hand the triumph of 'democracy' and, simultaneously, on the other the maximum number of 'home impacts' at a relatively modest cost per thousand per 30 seconds.

By the first quarter of 1957 TAM was able to report that the BBC's share of the national audience had fallen to a mere 26 per cent against ITV's 74 per cent.† If one took the lower

* In 1970 the meters were called SET-meters, and were managed by Audits of Great Britain Ltd. There were 2,650 of them.
† BBC Audience Research put the split at this time at BBC 33 per cent, ITV 67 per cent.

income groups the flight from the BBC appeared almost total. And it seemed clear that this was more than just a vote for or against particular programmes. It had a socio-political element: it was a vote against 'Them', the Officers and Gentlemen, the 'Toffee-nosed', the 'Highbrows', the 'Eggheads', whom the BBC now seemed to represent when set alongside the breezy, mid-Atlantic style of the ITV 'news-casters' (the BBC had only 'announcers'), the gregariousness of the London Palladium, or even the mateyness of OMO and Murray Mints, 'the-too-good-to-hurry mints'.

It was a sad comment on British society at this time that the dilemma posed by this rapid polarisation should have been real enough. It could be argued that in broadcasting terms it was Reithian élitism coming home to roost. However that might be, there seemed little clear alternative between a profit-propelled Numbers Game on one side and the 'effortless superiority' of a Platonic 'Guardianship' on the other, little cultivated public ground between the formal gardens of the Big House and the machine-made furrows of the market. The central issue of choice in an advanced, educated industrial democracy had been almost wholly ignored by a Labour Party until lately class-bound, and now lost between its fading Morrisian idealists and its up-and-coming technologist-economists, already firmly engaged in the larger Numbers Game.

Because of this, the moment was missed; without an alternative the Labour Party clearly could not redeem its pledge to reverse the commercial television decision. Before there had been a real chance to work out the problem, as postwar Britain *might* have done, the issue was statistically—and thus authoritatively—foreclosed.

The Other Inflationary Spiral

The British now became increasingly aware of what on the other side of the Atlantic Newton Minnow, chairman of the Federal Communications Commission, called 'the desperate compulsion of some of our licensees to work and plan and live by numbers.' And this was a compulsion which now also embraced the BBC for whom it was clearly not politically possible to stand by and watch its audience spirited away by commercial television's lowest-common-denominator strategy. Indeed from 1955 onwards, when the BBC began to publish its own head-counts, there grew up between the BBC and ITV something reminiscent of the ratings warfare of CBS and NBC. 'The daily chart comparing BBC and ITV audiences tended to become a, if not the, vital document in the determination of programme schedules,' reported E. G. Wedell,[12] in 1961 Secretary of the Independent Television Authority.

Competition, Reith had warned the Beveridge Committee on Broadcasting, would be 'competition in cheapness, not goodness'. It was not wholly true—for the complacency of the BBC was usefully shaken, causing it to greatly expand its news services and to open itself to greater public controversy and to the freer play of ideas. Yet it was true enough. In particular, competition brought structural changes, deeply bound up with value-making, which appeared as irreversible as in the newspaper field.

Mr Huw Wheldon, a BBC producer of that period, has vividly described the onset of the Law of Copycattery at the BBC. '... once you started losing audiences it was not only very difficult indeed to win them back, but extremely easy to lose more ... The brute fact is that pure competition on US lines forces companies into a deeply like-against-like situation and even in the more complex context such as obtains in the UK ... the drift is inescapably towards the same conclusion. And in turn this leads to a diminution of the spectrum or range of programming possibilities as a whole.'*

'The BBC,' wrote its historian, Asa Briggs, 'always remained outside the market complex.'[13] This had, indeed, been its great merit, and the ultimate source of the trust which so many people reposed in it. Whatever deficiencies they may have had, its values were honest professional, not market or monetary, values.

But its new rivals, the major commercial TV companies, were dominated by show-business impresarios, radio and electricity industry tycoons, newspaper interests, and financiers—men who thought in market terms of buying and merchandising

* Speech to the Royal Society of Arts, May 1971, when Mr Wheldon was Managing Director of BBC Television. Reported, the *Listener*, 13 May 1971.

talent. Their first step, echoing the war of Hearst and Pulitzer, was to try to buy the BBC's staff with vastly enhanced salaries. They succeeded, initially, with four hundred or so. Then, between 1956 and 1959 television advertising more than quadrupled. Money flooded into the company coffers. By the end of 1958 the *Daily Mail* was reporting that the initial £2,250 investment of Mr Norman Collins, the vice-chairman of Associated Television, who had discerned in commercial television one of the Four Freedoms, had grown to be worth £500,000. Meanwhile, at Scottish Television, the Canadian, Roy Thomson, who had learned his money-spinning techniques in Canadian small-town commercial radio, was employing what he disarmingly called 'a license to print money' to such effect that within two years he was able to use the proceeds and the company in a 'reverse takeover bid' which made him chief proprietor of the *Sunday Times*, of two other national Sunday papers, and a string of important provincial dailies and weeklies all over Britain.

By 1961 the incomes of the combined commercial companies were more than twice that of BBC television. To sustain its audience ratings the BBC was now obliged to bid competitively in the rising—indeed inflationary—market for talent; between 1955 and 1960 programme costs doubled.[14] Inevitably, at this level, audience rating and rate of reward began to form a basic equation: the yield on the investment. And, in turn, to ensure this, as the level of monetary 'risk' now required, there had to be a vast inflation of publicity, endless trailing of forthcoming programmes, endless gossip—particularly, and above all, money gossip. 'TILL DEATH US DO PART' WRITER WINS £40,000 2-YEAR BBC CONTRACT, ran a characteristic headline, referring to a comedy series of occasional brilliance and, perhaps more to the point, viewer and headline-pulling notoriety.

The BBC, for all its integrity and its record in the sustaining of high standards, was absorbed like all the rest into the febrile market—and market psychology—of the booming entertainment industry, which was soon to bring the floating of the Beatles and other performers on the London Stock Exchange. And if, at first, Northern Songs, the company holding rights in the compositions of three Beatles, was received with raised eyebrows in the City of London, these came down to more normal levels when the price of the company's shares quadrupled in three years.*

The higher costs rose in this somewhat desperate effort to increase, or at least hold, the ratings, the higher became the minimum 'economic' audience.

With a nervous eye on the charts, public and commercial rivals ran near-identical formulas back-to-back like weary boxers locked in a perpetual clinch; planners designed whole evenings to maintain the glazed inertia of the viewer and ward off the dangerous moment when he might reach out for the switch.

For the forces that will systematically gather the largest television audiences—and the *smallest* would be two or three millions†—and then hold them reliably anaesthetised are strictly limited. As the monitors clearly reveal, one of them is force of habit; the crease that stays in the paper. As Robert Silvey, the director of the BBC's audience research put it: 'In general, the longer a broadcast remains unchanged, the greater the number of its habitual listeners.'[4] (And no viewer in either Britain or America will need convincing of the faithfulness with which this law has been followed.)

And, along with the sweet—or deadly—familiarity that hypnotically holds, the novelty that attracts—but the sharp appearance, rather than the *reality*, of change—the cultural equivalent of the annual model styling changes of the 'efficient' car manufacturer. (As that master of advertising arts, David Ogilvy, says in his handbook, 'The two most popular words you can use in a headline are "free" and "new". You can seldom use "free", but you can almost always use "new" if you try hard enough.')‡

Hence, as the statistical mills ground on, there appeared that curious, insistent, vaguely worrying phenomenon of the

---

* Top performers in appreciation in the London Stock Market in 1969, Australian nickels excepted, were the shares of Management Agency and Music, owning the services of the singers Tom Jones and Engelbert Humperdinck. They quadrupled in nine months.

† This is for Britain. In the United States a night-time television programme enjoyed by only about ten million people, or six million households, is likely to be quickly taken off as a failure.

‡ David Ogilvy, *Confessions of an Advertising Man*, New York. Mr Ogilvy is famed for his success in making the Hathaway shirt 'new' by giving the bearded English model a black eyepatch.

sixties, endless, frantic change that somehow changed nothing, the phenomenon of 'the Top' that was always moving, yet always the same, a fixture in the headlines. How odd to reflect that 'the Top Ten'—in popular music—did not begin to appear in Britain until 1943 (published by the *Melody Maker*). Now the BBC and the commercial companies figured weekly in the statistical combat of the 'Top Twenty' programmes—not infrequently bickering bitterly over the accuracy of each other's figures. Right or wrong, the authority of the rating appeared total. In 1969 the chairman of the Independent Television Authority, Lord Aylestone, formerly Herbert Bowden and long the Labour Party's Chief Whip in the House of Commons, told an interviewer: 'I find it a little difficult to understand criticism of the programme content when currently we are getting 17 programmes out of the Top Twenty almost every week. Our audiences shares are 54s and 55s [per cent] ... our programmes are probably better than they've ever been.'[15]

How insidious these institutionalised numerical processes can be, and how far the inflation has, in fact, gone is well illustrated by a comment from a recent BBC director-general, Sir Hugh Greene: 'As broadcasters we can so easily be terrified into the thought that nobody is listening to us, nobody is watching us. And in broadcasting terms, "nobody" can often be numbered in hundreds of thousands, or even millions, if, on the other channel "somebody" amounts to ten millions.'[16]

In 1937 Robert Silvey, newly arrived at Broadcasting House to organise listener research under Reith's beetling brows, explained his purpose with the propitiatory words: '... the task of the helmsman will be made infinitely easier by good charts.'

By the 1960s, however, the helmsman appeared to be lashed to the wheel, and for much of the time revolving with it. It was the charts themselves, now computer-linked to an automatic pilot, which appeared for much of the time to be steering the ship.

Lifebuoys to Cling to

Time itself, the human living space, had now become a commodity, sliced wafer-thin, with a precise monetary value per slice. 'In television,' advises David Ogilvy, 'you have exactly 58 seconds to make your sale, and your client is paying £200 a second. Start selling in the first frame and never stop selling to the last ... The average American consumer is now being subjected to 10,000 commercials a year. Make sure that she knows the name of the product being advertised in your commercial. Repeat it, *ad nauseam*, throughout.

That was America; but in Britain, where the ITA had now ruled that a 'natural break' in the programme (to which the Act confined advertising) 'is not unnatural because it has been contemplated' the nausea level was now rising satisfactorily. Critics were frequently assured that the Authority had examined all advertising to ensure its truth and good taste. If Jesting Pilate had stayed for an answer he would undoubtedly have received a well chosen one, but the fact remains that one does not sell competing mass consumer goods, many of which are virtually identical but for their brand names and labels, by appeals to objectivity and reason. The aim indeed must be the swiftest possible suspension of the critical faculties; and with their catchy jingles, verbal mumbo-jumbo, matey jokeyness, sweet slavering infants, roared injunctions, whispered incantations, discreet preoccupations with stains, and plainly venal interviews with complacent suburban housewives, the commercials increasingly provided the cement of a world where nothing any longer was entirely true and nothing was entirely false.

'There is a place for endless forms of fantasy, indeed some kind of real need for them,' writes the American Jesuit moralist William F. Lynch in his study of mass communications. 'But we must always be able to know when we are playing and when we are dreaming; we even have the right to be stupid and tawdry, provided there is the slightest element of the deliberate or conscious judgment in the process. For judgment is all; when it begins to go, all culture is moving into great danger.'[18]

Alas, in the enveloping miasma, the endless 'impacts', individual judgment, battered and baffled, was increasingly replaced by the mechanised, mathematically-expressed 'score' of 'scientific' authority and intimidating precision.

And yet—and here was one of the central and most fatal aspects of the Numbers Game which we shall encounter in many of its fields from economics to opinion-polling—every intelligent person who knew anything of the subject was well aware of the awful hollowness of the ratings on which a new world of values was ostensibly being erected. In both Britain and America doorstep checks have shown that the number of sets switched on, and thus registered by the automatic meters, may be an unreliable guide to the numbers of people actually watching. One investigation, covering 11,400 housewives, undertaken by the J. Walter Thompson Advertising agency in Britain in 1961, showed that between 8.30 and 9 p.m. in homes where sets were switched on, only half of the women were in the TV room and paying full attention, while more than 12 per cent were not watching at all.[19] Hourly viewing logs kept by sample viewers are supposed to make possible adjustments for such deficiencies in the instruments, but the reliability of ordinary human beings performing this somewhat absurd chore is open to doubt. In the United States, where a considerable number of rating services operate, the divergences have been so bewildering that a Radio Television Ratings Review was established by the Advertising Research Foundation to resolve them. Its findings, issued after a year's investigation, were violently disputed.

It hardly matters. For the more pervasive the uncertainties, the more eroded the judgment, the more desperate the psychological need for a few, clear-cut figures to hold on to. That is the peculiar irony of the thing. For it is the fragmentation, the multiplying distortions, the mechanical agitations of the Numbers Game which themselves induce the incoherence and loss of steady judgment which makes its crisp numerical 'indicators' so indispensable—lifebuoys at which the drowning man clutches in desperation as they float by. The Numbers Game is self-feeding and self-perpetuating.

This particular vicious circle, a basic mechanism of the Numbers Game, is already well advanced. Across the broken ice of sexual morals, Dr Kinsey's Institute for Sex Research—and later emulating surveyors—stretch their endless statistical ladder (without noticeably increasing the sense of security). In the morass of print, more and more papers offer the certified stepping stones of the monthly best-seller lists. 'Best-seller' lists had appeared in the American *Bookman*, it is true, since the turn of the century, but now they were becoming an institution of the whole 'Free World', a statistical artefact buttressed and gilded by monetary values, by endless gossip of prices paid by film companies, or titanic deals with periodicals and reprint houses. Only half joking, Harold Robbins, of *Carpetbaggers* fame, intimated to a British television interviewer that he, Robbins, was 'the best writer in the world'. Because with sales and advances like that, he just had to be. It was a form of logic increasingly hard to resist. True, neither painters nor sculptors yet had their Net Sales Certificates. But since 1968 the London *Times* had offered, calculated on the logarithmic scale, the rising curve of fine art prices, the *Times–Sotheby* index which showed Picasso up $10\frac{1}{2}$ between 1960 and 1969, Lautrec up six times, Whistler $4\frac{1}{2}$ times. The younger artists had not yet achieved their charts, although if they had, their peaks and valleys would certainly have been as precipitous and almost as authoritative as those of most 'pop' singers.

Curves on a graph do not yield absolutes in the old sense. Yet trends, while rising or falling, are even more tyrannical. And the trend is the characteristic social formulation—the basic ideology—of a statistics-dominated age. Stuart Chase has explained that one of the services of the social sciences is to alert us to trends so that we shall not be so ill-advised as to try to move against them. Indeed, as any stockbroker will tell you, there isn't much you can do about a trend, except get on—or out —at the proper moment, and the widening application of this observation has been vividly demonstrated in the last two decades or so by the remarkable facility with which the art critics have turned from, say, the interpretation of Action Painting (just how many trends ago was *that*?) to the message of Op Art or the vital truth of Pop Art. They have triumphantly survived the heavy drafts on their funds of enthusiasm, and the breadth and depth of their understanding has lent a new fascination to the pages of the 'quality' newspapers.

But some sort of an ultimate must surely have been reached with Andy Warhol's claim to have dispensed with both technique and art in favour of the Thing Itself, a proposition surely brilliantly made out by the stack of Brillo cartons he recently exhibited at the Tate Gallery, but on view less expensively and exactingly at any large supermarket.

Warhol, a young man of remarkable insights, has said that 'in future everyone will be famous for at least fifteen minutes.' He has also declared—according to *Time* magazine, 'with utter sincerity'—'I want to be a machine.' This, too, has the appearance of a logical conclusion. Unfortunately, for those of us not yet able to share this wish—or Warhol's apparent unconcern about the machine's gearing and power source—it seems only too likely to be gratified as the automated processess of the Numbers Game penetrate larger and larger areas of human activity.

'Democracy to Complete Itself ...'

In the minds of most people in the United States or Western Europe democracy is simply defined. It means the fundamental principle of 'one man, one vote', and it means that when the votes are counted, the numerical majority shall prevail.

With continuous head-counts in progress, the majority's rightness, hitherto only insisted upon at election times, was now underlined and documented daily, and further confirmed by the vast sums of money dancing attendance upon it. Perhaps more important, anything less than the majority became—not merely members of the community with different opinions—but that suspect thing, a *minority*.

Thus, if thanks to the vitality of the Reithian tradition, the BBC still put out 'low rating' programmes, their 'minority' status tended to be 'licensed' and their future precarious.* In 1957, after a loss of audiences to commercial television and Radio Luxembourg, the BBC's Light Programme, the broad base of Sir William Haley's famous aspiring pyramid, was made lighter still by cutting out 'audience-losing' serious features and music. Its role was now officially described as to be 'undemanding ... its guiding principle that it shall never cease to entertain ... with comedy, light music, and light drama.' At the same time, the 'minority' Third Programme was truncated.

---

*Thus, in his speech on BBC policy and the effects of commercial competition, delivered to the Royal Society of Arts in May 1971, the managing director of BBC Television, Mr Huw Wheldon, interestingly described the

The BBC, protested the Sound Broadcasting Society, which numbered T. S. Eliot among its supporters, was surrendering to 'the Mass Audience drifting aimlessly on its tide of treacle'. But such 'minority' protests were increasingly discounted. For—as an incident in the mid-sixties showed—the 'electronic' vote transmitted by the radio and TV monitors was now visibly impinging on the old British political system, and was beginning to transform it.

At this time a number of 'pirate' radio ships had anchored off the coast of Britain. They were in contravention both of international wavelength conventions and of copyright law, and they were vigorously thumbing their noses at Britain's long-standing prohibition of radio advertising. Rattling off endless 'top of the pops' records, advertising spiels, and disc-jockey chats, they became something of a cult with teenagers, partly because of the ships' 'piratical' status.

As any Government would, the British Government wished to close the pirates down. But one ship claimed ten million listeners, another, nine.* 'The Government,' commented *The Times*, 'have been reminded that if they value the good opinions of young voters they had better make other arrangements for providing the same sort of noise.'

Her Majesty's Government capitulated. In one of its rare direct interferences with the BBC, the Government instructed the Corporation to allocate one of its scarce middle wavebands to provide for 18 hours a day a non-stop stream of pop music, broken only by the chat of a string of disc-jockeys, whose manner, explained the BBC Controller, would be 'both professional and personal'[20]—a pretty 'professional' (if not 'personal') description of the sort of phoney intimacy involved. Meanwhile, the recently lightened Light Programme would go on being 'undemanding' on another wavelength, with sweeter

---

decision of planning staff in a two-channel operation to put on a programme likely to appeal to the smaller audience as 'an act of self-abnegation'—although he was in favour of this self-sacrifice being made when, and if, strategically possible.

* In characteristic Numbers Game fashion these intimidating statistics seem, in fact, to have been dubious. BBC Audience Research retorted that 77 per cent of listeners over 15 never, or hardly ever, listened to the pirates. Only 16 per cent listened nearly every day, and of these one-third were under 24 years of age.

popular (as opposed to pop) music and mellower chat.

An embittered educationist, S. Gorley Putt, wrote to *The Times*:

> At the very moment when the Labour Government may contemplate in the Welfare State the victory of one great humanitarian campaign, it proposes to concede defeat on the other battlefront by deliberately deluging its educational paupers in a ceaseless flow of contemptuous rubbish. Having risen above the reactionary parrot-cries of 'Give 'em a bath and they'll only keep coal in it' we are now to shrug shoulders and say: 'Give them a wireless set and all they want is drivel.'

POP GOES THE BBC, ran the headline above *The Times* leader on the changes. The note of finality seemed appropriate. The Reithian Principle of Justification by Faith had been undermined and finally driven from the field by the Stantonian Principle of Justification by Numbers.

'The BBC,' summed up *The Times* in the summer of 1969, 'responded to the challenge of commercial television by setting out to prove that they too could produce programmes for the mass audience ... To such an extent have they succeeded ... that they have intensified the tyranny of the audience-ratings, not only for themselves, but also for the independent companies.'*

The news columns with their endless bickering over audience figures bore out this verdict. Yet if the BBC now responded to what the newspaper magnate, Lord Thomson, a past master of the media Numbers Game, has called 'the happiness of being No. 1,' the BBC tradition was not wholly lost: with rich irony it had been entrusted to the computers, encoded, and incorporated into the Numbers Game. In a speech at Edinburgh University in March 1971, the Director-General of the BBC, Mr Charles Curran, explained that if the BBC's share of the British audience, as against that of its commercial rival, rose *above* the 50:50 level 'the BBC would have to look at the balance of the programming, because that would be a clear indication that we might be leaning too far towards the purely enter-

* Cf. also Miss Stella Richmond, programme controller, London Weekend Television in 1970: 'This ratings thing with the BBC has become a terrible game ...' interview, *Financial Times*, 19 June 1970.

taining. I should be similarly concerned if we fell below the 40:60 proportion, because that would mean that we were either not good enough in the quality of what we were offering, or too highbrow in our general balance to retain our hold on the audience.'[21]

A Numbers Game indeed, and of quite peculiar delicacy. But it is not only culturally that one touch of the Numbers Game makes the whole world kin. It becomes as comprehensive and enclosing as the feudal system itself, as its confident, mechanical and increasingly automated processes tighten around our thinking, transform our political life, and rule our morals no less than our economics.

5

CALCULATED OPINION

The present system is not democratic at all. It does not represent the *demos*, that is to say, the population taken as a political
entity, but the *laos*, the human particles of the mass accumulated without organisation or collective consciousness.
Salvador de Madariaga, *Democracy versus Liberty*.

Like other valuable articles, public opinion is surrounded by
many counterfeits.
Lord Bryce, *The American Commonwealth*.

On 25 July 1969 at Dukes County Courthouse, Edgartown,
Massachusetts, Senator Edward M. Kennedy was given a two-
month suspended sentence of imprisonment after pleading
guilty to leaving the scene of the accident in which Mary Jo
Kopechne lost her life when the car the senator was driving
plunged off a bridge on Chappaquiddick Island. Kennedy's
driving licence was withdrawn and he was placed under the
jurisdiction of the court for one year. The local police chief
pronounced himself satisfied, and the senator made a contrite,
if somewhat evasive, statement on television in which he invited his constituents' advice and prayers.

A fortnight later, *Time* magazine came out with its Louis
Harris opinion poll: 44 per cent of Americans, it reported,
believed that Kennedy had 'failed to tell the real truth'. And
if 38 per cent believed that the senator was 'not driving under
the influence of alcohol' as he claimed, 32 per cent believed that
he was. From then on, as scientifically sampled, scientifically
questioned, and scientifically computed by the Louis Harris
organisation, 'Americans' went on to express, item by item,
their degrees of belief, or disbelief, in the Senator's account of
that tragic July night. Then, moving smoothly from the merely
legal to the moral sphere, 38 per cent 'agreed' that after the
accident Kennedy 'seemed more interested in his political future
than in what happened to Mary Jo Kopechne'. However, re-
calling a well-known Christian injunction—or possibly just

facing the necessity of responding to yet another question—68 per cent felt it 'unfair to be critical of the way he reacted to the accident because the same thing could have happened to anyone'.

While *Time* was thus drawing on the wells of the people's wisdom, the question of further legal proceedings was still open. The Massachusetts District Attorney was in process of applying to a superior court for a formal inquest on the dead girl. It was November before procedure had been agreed and arrangements for the inquest set up.

On 7 November *Time* came out with a new Louis Harris poll in which the people's verdicts of August on Kennedy's conduct were neatly tabulated alongside the afterthoughts of October. Now only 32 per cent agreed that 'nothing immoral had taken place between Kennedy and the girl', a substantial drop, the statisticians dutifully pointed out, from the earlier 51 per cent. However, only 17 per cent went so far as to say that something had. The number feeling that the senator was not driving under the influence of alcohol had also fallen; and turning now from the judicial to the divine function, only 57 per cent 'of Americans' were prepared to agree that Kennedy had 'suffered enough'. One-third disagreed, presumably feeling that he ought to suffer some more. The inquest then proceeded.

We do not know what degree of attention or scruple each of the 1600 representative Americans pinned down by the Harris interviewers—according to the sampling prescriptions—brought to the lengthy and exacting questionnaire that was put before them. Certainly, in the midst of their various preoccupations, it must have varied widely. Nor have we any means of knowing the nuances of interpretation each brought to the questions, nor the range of meaning implied in the answers. We do know that on many points in this complicated and much obfuscated affair they—surely the largest jury in history—lacked both the factual knowledge and the basis of personal observation on which sound opinion can be formed. Unlike the old jury of twelve, picked by the crudest of pre-scientific sampling procedures, they had the advantage neither of studying the witnesses at close quarters under systematic cross-examination, nor of the clarifications and summing up of the judge, nor of discussion in the jury room with the sense of social responsibility there engendered. Each 'representative' American uttered his yea or nay or maybe in an isolation relieved only by the brouhaha of the mass media. And yet, from this far-flung sargasso sea of ignorance, half-knowledge, variegated prejudice and infinite ambiguity there somehow emerged, as by scientific immaculate conception, 'Public Opinion' of almost frightening solidity, a verdict of intimidating, decimal-point precision.

Down the years a vast weight of obloquy has been heaped on the heads of those devout, and sometimes learned, judges of Massachusetts who condemned to death the so-called witches of Salem in 1692. Yet they did so after much examination of the witnesses and the accused, after much earnest consideration of the evidence and searching of hearts.[1] Probably they would be both perplexed and outraged at the flippancy of moral judgment upon an individual arrived at by random or quota sampling, punch card and IBM machines, in the Massachusetts of the mid-twentieth century.

That trial by opinion poll should now be almost as readily accepted as once was trial by ordeal is indeed striking evidence of the extent to which this bizarre, yet characteristic, institution of the Numbers Game has become part of the fabric of our lives and societies. 'The polls', in their insistent, nagging, headline-hogging way, have tended to become moral, as well as political, social and cultural arbiters. In Britain, for instance, they played their part in such deep-going moral issues as the reform of the laws on homosexual conduct, on abortion, obscenity and capital punishment. In America even so traditionally authoritarian a body as the Roman Catholic Church conducted an opinion poll of priests on issues such as priestly celibacy and contraception (a majority disagreed with Pope Paul's condemnation).[2] On a central issue such as Vietnam, the polls tended to become a kind of surrogate of the national conscience. Harold Wilson has related that when he told President Johnson about British protests against the course of the war in Vietnam, Johnson replied: 'Have you seen what the polls are saying?'[3] It was a message which American Presidents would find increasingly compulsive.

Yet if the polls were indeed becoming a substitute national conscience, there were already signs that under the impact of their incessant, mechanical activity that organ was becoming increasingly engorged and flabby. When in the summer of 1971 Britain faced the momentous decision of whether or not to 'join Europe' by entering the Common Market, the *Daily Mail* observed that the verdict of the opinion polls would be critical. 'The next few weeks will tell,' it noted as the hour of decision in Parliament drew near. 'All eyes will follow the curve of the opinion polls.'

If all eyes did, what they saw seems to have had remarkably little effect on the Government. In October when the House of Commons voted for entry by a majority of 112, four polls were still showing half the British public *against*, only around one-third in favour. Despite this sustained hostility, 86 per cent of the people believed in September that the Government would nevertheless take the nation into Europe. No doubt, by this time, they were beginning to understand the rules of the Game; to suspect that the political parties scrutinised the opinion polls much as commercial concerns look at market research report as indicating the opportunities which it might, or might not, suit them to exploit. This one didn't suit Mr Heath, who contrived to remain remarkably unmoved by 'the verdict of the People'; it did, however, suit Mr Harold Wilson and may have had a good deal to do with the vigour and discipline of official Labour's *volte face* on the issue.

The Pulse of Democracy

John Locke described Opinion as one of the three categories of Law; and Rousseau, who seems to have coined today's ringing phrase (*l'opinion publique*), pointed out that even despotism needed to rest, in the last resort, on public opinion.[4] But as the practical Bryce pointed out in *The American Commonwealth* the difficulty about government by public opinion has always been 'the difficulty of ascertaining it'. All Walter Lippman had been able to discover in his book of 1922 on the subject had been a 'phantom'.[5]

That in an increasing area of Western society 'Public Opinion', scientifically analysed and precisely measured, is now more or less permanently on tap is a major development from statistical science whose implications have even now not yet been adequately considered.

'Not only have the polls demonstrated by their accuracy that public opinion can be measured,' wrote Dr George Gallup in his classic work, *The Pulse of Democracy* (1940), 'there is a growing conviction that public opinion *must* be measured.' If the New England Town Meeting was for the most part no longer practicable, the public opinion poll had appeared as a timely substitute, at once clarifying and educating as it posed the key questions in 'language understandable by the great mass of people'. As for those who still argued the outdated issue of Representative versus Direct Democracy, wrote Gallup, in their unwillingness to trust the common people they differed 'only in degree, not in essence, from the view urged by Mussolini and Hitler that the people are "ballot cattle".'

Certainly, as an addition to the apparatus of 'checks and balances' the opinion poll can today have real value—if only in divesting the many vested interests and lobbyists of those sacred robes of Public Opinion which do not rightly belong to them. In Britain the polls emboldened the Minister of Transport in bringing in 'breathalyser' tests—greatly cutting drunken driving—despite the efforts of the brewers' lobby to raise the potent cry of Freedom.[6] In fact, both in Britain and the United States the polls have on occasion done invaluable service in showing that the public was more mature and realistic than wary politicians allowed. Although in 1957 the Wolfenden Committee reported in Britain in favour of removing homosexual behaviour between consenting adults in private from the sphere of the criminal law, to legislate on these lines was then considered tantamount to political suicide. Action was delayed for ten years—until opinion polls showed MPs' fears of constituency reactions ill-based. Indeed, Lord Arran, initiator of the liberalising Bill in the House of Lords, has recently described the 1965 verdict of the National Opinion Poll on the issue as his 'trump card' in the long, often bitter and uncertain Parliamentary fight for a more tolerant approach by the law. In the same way, the polls strengthened the Government in reforming the abortion law, another proposal which not long before would have been considered unthinkable.[7] Again, at an earlier date in America, it was the polls that brought the 'forbidden' subject of syphilis out into the open, making possible the setting up of public information centres.

It is perhaps in this way, as a form of participation in actual policy and planning, that opinion or attitude surveys display their most solid worth. But this quiet, purposive use is a far cry from the quick-fire opinion polls we see most of the time, a booming sector of a booming mass media industry, feeding into the headlines an endless conveyer belt of Instant Opinion, aspiring to become prefabricated history, yet in fact the corollary of the audience rating or position in the Top Ten.

Dr Gallup in his book points to 'the fundamental right of every person to give expression to the worth that is in him', and he justly emphasises the 'grave danger to democracy which arises when the ordinary citizen begins to say: "It doesn't matter what I think—they'll run things their own way anyhow." ' Dr Gallup sees opinion polling as a way to ward off this creeping paralysis of our time.

Unfortunately, a generation of proliferating opinion polls suggests the contrary. The individual is likely to find minimal satisfaction as a code-mark on the clip-board lady's questionnaire or an integer in opinion calculations; alienation is more likely to be confirmed than relieved by these complex, impersonal and specialist processes.

Ignorance Transfigured While You Wait

If the old rhetoric of Public Opinion was hollow, the precision of the 'continuous audit' or daily opinion barometer is more subtly dangerous.

In the conventional Western democracy a citizen votes only at fairly infrequent intervals, and then not on an infinite range of complex issues, but for or against a party. Those who neither know nor care tend to absent themselves from the polls. By contrast, the opinion poll in effect institutes the compulsory vote—for the well-trained interviewer is known by his or her small proportion of refusals from those upon whom the 'sampling' lot falls. Hardly remarkable therefore that there is by now a formidable accumulation of evidence that the precise figures of opinion 'audit' contain much counterfeit.

After research on the data of the Gallup, Fortune and other polls, H. H. Hyman and Paul B. Sheatsley, of the National Opinion Research Centre at Columbia University, concluded: 'Although tests of information invariably show at least 20 per cent of the public totally uninformed (and usually the figure is closer to 40 per cent), the 'no opinion' vote on any poll question seldom exceeds 15 per cent and is often much lower ...'[8] To admit ignorance must go against the grain for most people. When Britons were asked in a National Opinion Poll in 1967 whether the decision to devalue the pound sterling had been right or wrong, only 12 per cent of those interviewed failed to pronounce on this intricate and vastly technical question. Yet in 1959 20 per cent of the electorate was unable to name a single party leader, and 28 per cent even of those who claimed to be party members were unable to name even three of their own party leaders.[9]

It has, of course, also been demonstrated *ad nauseam* that different poll interviewers, merely by virtue of appearance, personal chemistry or unvoiced expectation may get radically different results from the same people; or the same interviewer will get different answers merely by changing the sequence of the questions, or by some insignificant variation of the wording. In a follow-up after a regular survey, workers of the London School of Economics Survey Research Centre concluded in a 1967 report that 'it was unusual for as many as half the respondents who answered the question to have interpreted it precisely as intended.'

Even where true opinion exists, and is strongly held, it by no means follows that it is going to be revealed to a total stranger on the doorstep. Or it may be that the opinion, while genuine, and indeed deep-going, may never have been articulated or even fully realised. 'People say one thing, they often mean something else,' comments Conrad Jamieson, American principal of a British social survey research firm.[10] 'The "private" and "public" opinion of a person seldom agree,' adds Gunnar Myrdal. It is too easy to dismiss the questioner with the conventional 'correct' view. The classic instance here is the American poll where only 1 per cent said 'No' when asked whether they believed in the Freedom of the Press. But when later questions asked about freedom of publication for socialists, then it emerged that only one in four of these upholders of the Freedom of the Press would permit the Socialist Party to publish newspapers, and that one in three were in favour of banning newspapers which criticised the American form of Government.

On a good many matters, even of basic principle, 'Public Opinion' as registered by the polls shows large and erratic swings from year to year. From a poll verdict showing 62 per cent in favour of Britain's joining the European Common Market in mid-1966 'opinion' had switched to 51 per cent against in 1967 and 64 per cent against in 1970. In 1947 a Gallup Poll showed 68 per cent of Britons in favour of hanging for murder. In 1955 another poll showed 65 per cent *against* hanging; but by 1969 this had become over 80 per cent *in favour of* hanging.

Do such changes represent true changes of opinion after private discussion and public debate or valid changes in the evidence? The overall murder situation had, in fact, changed little, with the rate remaining low,* despite the fact that hanging had been largely abolished in 1957 and completely abolished in 1964.[11] But one or two particularly brutal murders had filled the headlines, which also stressed the general rise in crimes of violence.

Far from being a re-creation in modern form of the old New England-style Town Meeting what we have here is much more like an odourless, bloodless, and indeed almost noiseless equivalent of the old street mob, turned hither and thither, no longer certainly by rabble-rouser and a 'cry', but by TV sensation, by the impact of current headlines, and the morbid effects of prolonged statistical addiction.

This may be an over-generous view. Recent research suggests that—to a degree that would generally be crucial—the precise, immaculately conceived verdicts of opinion polls represent, more than anything, the operation of pure chance. P. E. Converse and his associates of the Opinion Research Centre at Michigan repeated an opinion survey of the same random sample of electors on a basic, semi-ideological issue three times —in 1956, 1958, and 1960. Although all respondents had been

---

* The total of murders in all Britain was less than one quarter of the figure for the New York Metropolitan area alone.

'filtered' by first being asked whether they in fact had an opinion on the topic, an analysis nevertheless suggested that only one-fifth to one-third of the replies represented meaningful opinion. For the rest, answers varied so wildly both over time, and in comparison with respondents' other views and circumstances, that the researchers were driven to the conclusion that they were, in effect, made at random.[6] But perhaps random isn't perhaps quite the right word, because psychological research has shown that there are people who like to say 'Yes' and people who like to say 'No'—so that by changing the question from the positive to negative form one changes the response.[12]

In Britain a repeated opinion survey on nationalisation, carried out by David Butler and Donald Stokes in 1963 and 1966, suggested broadly similar conclusions (with only 39 per cent in this case holding to any identifiable view).[13]

Measuring the Immeasurable

'The ideal director of a public opinion poll,' Dr Gallup has written, needs to have 'laboratory objectivity' but at the same time 'always needs to be "on top of the news".'

If the authority of the opinion poll derives from the white coat of science which the pollsters wear, no one can accuse them, in America at least, of failing to keep it freshly laundered. In numerous institutes and 'centres' of opinion research prodigious pains are taken in the pre-testing of the wording and order of questions, in the development of attitude and intensity scales, of 'multiple-choice' and structured and filter questions, and—to quote the title of one two-year study—'The isolation, measurement, and control of Interviewer Effect.' Thus the researchers of the University of Denver Public Opinion Research Centre, having first demonstrated that four out of five respondents to a certain question were either lying in their teeth or suffering from amnesia, did not despair. For once detected, they tell us, invalidity too 'can be measured and analysed' and controlled.[14]

It is heroic—yet at the last moment the human specimen wriggles away from the sociological pin.

After extending possible reactions from three to five the

Office of Radio Research at Columbia University reported that 'the use of a five-point scale tends to decrease somewhat the number of Indifference reactions.'[15] Very likely; but does it decrease the amount of actual indifference? With all the most sensitive scalograms and scalometers in the world, how can one be sure that one man's 'very strongly' is really different from another man's 'quite strongly' or even 'not very strongly?'

Again, an 'open-ended' question may reveal something of the true complexity, the evanescence, the sublime quirkiness of the human mind in the process of generating an opinion. But the problem is merely moved a stage further on—to the classification, coding and quantification of the shifting web of responses. For as Professor Hans Zeisel insists in his instructional work, *Say It With Figures*, 'the classification must be logically correct, the categories must be mutually exclusive, with no overlaps, and yet must include all possibilities. Each item must fit into one group and only into one group.'

C'est magnifique, mais ce n'est pas la vie!

Professor Hadley Cantril of Princeton, that Mecca of the polling order, has set out no fewer than eleven types of poor questions which should be avoided by scientific pollsters.[16] Yet again and again any normally scrupulous person will find that the only honest answer to an opinion poll question will be 'it depends,' or even, 'it depends just what you mean.'

With his quantifying preoccupations, the professional pollster may develop curious blind spots. After commissioning an opinion survey, the recent Congressional Commission on Obscenity and Pornography announced that contrary to the view of Congress which requested the investigation of this 'matter of national concern' the prevalence of pornography in America was not in reality a matter of public concern at all. The question which gave this result ran as follows: 'Would you please tell me what you think are the two or three most serious problems facing the country today?' Only 2 per cent of the respondents mentioned pornography or, as the report puts it, 'erotic materials'.[17]

Considering the number, urgency, and catastrophic character of the problems in fact pressing on America today—Vietnam, the race conflict, violence in the cities, drugs, youth rebellion and so on, this failure to single out pornography was hardly

remarkable. What was remarkable was the conclusion drawn.*
It affords a fascinating glimpse of the inroads of statistical addiction, with its assumption that human concern is somehow
strictly finite, existing on a single level, its numbered portions
interchangeable like automobile parts. To that fine, and comprehensive, Quaker phrase, 'We have a concern,' these surveyors would doubtless retort: 'You can't. You've exhausted
your quota.'

The greatest danger comes with the next stage when the
clear verdict of the People appears in the headlines under the
double warranty of Democracy and Science. Standing forth in
public print, its 'objectivity' is further reinforced: it becomes a
statistical artefact. How can people *be* concerned when it is a
'fact' that they are *not* concerned? The question is foreclosed.

In an instance as crude as the '2 per cent concern' cited, this
may not happen. But the onset of statistical pre-emption may
be insidious, the change of attitude gradual and apparently
'natural', and yet the whole 'objectivising' process may be consciously employed for political ends—as was vividly illustrated
in an issue which developed in the English Midlands in 1967-
1968.

Build-up by Numbers

No area of opinion is more complex, highly sensitive, and
delicately balanced than that of 'race relations'. In street or
workplace the neighbour with the strange accent or different
pigmentation may appear as a human individual—or as a less-
than-human creature, nourishing all manner of evil designs.
How the relationship develops will depend to a high degree on
the surrounding climate of opinion.

In Britain in the fifties the auguries were, on the whole,
good. Until 1962 British Governments continued, perhaps un-
realistically, yet sincerely, to uphold the doctrine that any
citizen of the Commonwealth, of whatever origin and skin
colour, must enjoy the right to be admitted freely to 'the

---

* In fact, a Harris Poll in 1969 reported that 76 per cent wanted pornographic literature outlawed and a Gallup Poll in the same year that 85 per
cent favoured stricter laws in this field.

Mother Country'. And if, in 1962, the Commonwealth Immigration Act for the first time imposed restrictions, it still gave automatic right of entry to the spouses and dependants of immigrants already in Britain; and three years later the Race Relations Act outlawed racial discrimination and incitement to racial hatred.

Both Labour and Liberal parties had passionately opposed the 1962 Control of Entry Act; but even after this major shift, the two main parties refrained from making an election issue out of immigration, moving towards a consensus based on humane control on the one hand, and improved social provision and community relations on the other. In the autumn of 1967 the first Race Relations Board, with seven conciliation committees, started work; and in 1966 the Home Secretary had extended the anti-discrimination Act to cover discrimination in housing, jobs, and in credit and insurance facilities.

In March, 1968, the Gallup Poll announced that 72 per cent of Britons approved 'the measures the Government is taking to control immigration'; and in early April, 42 per cent approved the anti-discrimination legislation. It seemed possible that with her wealth of community and social services, Britain might succeed in absorbing a coloured population which still amounted, overall, to only 2 per cent of her population without evoking the spectre of race hatred then so distressingly visible in the American South.

Generous, bold and imaginative leadership might be required; what was *not* required was politicians playing the Numbers Game, limply scrutinising successive opinion polls on the subject and calculating costs in votes at the next election. This, unhappily was what Britain largely got.

For on Saturday 20 April 1968, at the Midland Hotel, Birmingham, Mr Enoch Powell, a member of the Conservative Shadow Cabinet, delivered a speech, exercising, as he hinted, 'the supreme function of statesmanship ... to provide against preventable evils.' He retailed, with sympathy, the tale of a 'decent, ordinary fellow Englishman' who wanted to leave England because, he complained, 'in fifteen or twenty years time the black man will have the whip hand over the white man.' By dint of statistical extrapolation worthy of Malthus himself Powell built up an 'objective' picture of this takeover by coloured immigrants, who were indeed already making Englishmen 'strangers in their own country'.*

Official Conservative policy had suggested assisted voluntary repatriation merely as one aspect of a total programme; Powell now put it boldly in the centre of what the *Daily Mail* was soon headlining as POWELL'S SEND THEM HOME RACE PLAN.

As for the new anti-discrimination Act, this was to 'risk throwing a match into the gunpowder'. It seemed, however, that the gunpowder was still a little thin on the ground, for Powell now went on to spread more around by reading out a letter from a woman old age pensioner, a widow, whose home had been outflanked by blacks, and who was now being terrorised by them, with 'excreta pushed through her letter-box' and 'wide-grinning piccaninnies' yelling 'Racialist' after her in the streets. 'As I look ahead,' concluded the statesmanlike Powell, 'I am filled with foreboding. Like the Romans, I see the "River Tiber foaming with much blood."'

The speech achieved blanket coverage, which was further multiplied when the Conservative leader, Edward Heath, summoned Powell, and dismissed him from the Shadow Cabinet. But now the polls entered the picture and the headlines. The *Daily Mail*'s National Opinion Poll showed that 61 per cent of the British people believed that Heath was wrong to dismiss Powell. 67 per cent agreed with the views he expressed, a proportion rising to 70 per cent among Conservatives, in the skilled working class, and in the Midlands in general.

Later that year, a social research group checking on this support found that of those who had indicated approval of Powell's policy, less than half were accurate in their description of what this in fact was.[18] This, as we have seen, would not be unusual. However, statistically solid and incontrovertible in the headlines, the 'verdict of the British people' inevitably had its effect. Only a few weeks earlier 72 per cent had registered approval of the Government's immigration measures. By late April approval even for the anti-discrimination law, as calculated by the Gallup Poll, had dropped to 30 per cent.

---

* Special factors (e.g. age) contributed to high immigrant birth-rates and there was no reason to believe these would continue indefinitely. Indeed official Birmingham figures issued in 1970 showed West Indian and Jamaican births declining each year since 1963.

From this new base in 'public opinion'* a now 'vindicated' Powell was able to proceed to further headline-worthy revelations and further impressive statistical extrapolations, foreshadowing, as he said, 'reproducing in "England's green and pleasant land" the haunting tragedy of the United States.' In the Eastbourne speech from which this phrase comes, Powell saw the once proud Briton as 'the toad beneath the harrow'. He proposed a Ministry of Repatriation to speed the return of the immigrants.

Outlined in statistics, both his own and the polls, as in neon lights, Powell was now instant news in himself, his every public utterance a first-class pseudo-event. A fortnight after his Eastbourne speech, replete with horror stories, he was featured on the peak BBC serious programme 'Panorama' which—the approach having, it seems, become almost obligatory—at the same time gave the results of yet another opinion poll which it had itself commissioned from the Opinion Research Centre. This poll was, in fact, two polls, neatly and no doubt fittingly, segregated by skin colour.

When the whites, a nation-wide, allegedly representative sample of 520, were invited to say whether they thought the Government should stop the wives and children of coloured immigrants already resident in Britain from joining them, 35 per cent in their kindly little hearts said they did. 74 per cent agreed with Powell's 'large-scale plan for the voluntary repatriation of immigrants' and 35 per cent favoured compulsory repatriation, which the question briskly clarified as 'i.e. deportation'.

When the coloured 'sample'—420 in twenty areas—were asked, among other things, whether they would like to return 'to your country of origin if you received financial help', 38 per cent said 'Yes' and 43 per cent 'No.' 8 per cent had a little difficulty with the question since Britain *was* their country of origin; and, hardly surprisingly, 11 per cent didn't know.

Every working journalist is familiar with the technique of

---

* Cf. The Institute of Race Relations Report, *Colour and Citizenship*: 'The influence of public opinion, both as projected through the mass media and reflected in the opinion polls, grew through this period, and its relevance was eventually recognised, and then exaggerated.'

asking a question slanted in the desired, news-making direction, getting a hesitant or careless assent, or merely a vaguely affirmative grunt, and translating this into a bold affirmative statement. But it has taken the *scientific* opinion polls to erect this dubious gambit into a 'democratic' institution, adding only the rarely mentioned caveat of 'sampling error'. In the particular case of the coloured citizens in the 'Panorama'-commissioned opinion poll, perhaps hard-up, chilled by the English winter, and suddenly assailed on the street by some strange middle-class woman with a clipboard and a lot of questions, the suggestion of a trip home, *plus* an unspecified amount of cash must surely have powerfully invited at least a mumbled assent. Equally certainly, it could hardly be taken as a considered statement of true intention. In this case indeed, the polling organisation, with the customary scientific detachment, did point out that 'significant numbers had refused to be interviewed' and that the results should therefore be used with caution.

With what caution any poll results are likely to be used, particularly if they appeal to emotion, we have seen. Diligently building up on the 'Panorama' poll, as the poll was built up on him, Powell now asserted in another statesmanlike speech at Wolverhampton that 47 per cent of coloured immigrants, 'or practically half, opted for repatriation' and he now therefore proposed a £300m repatriation plan which would be welcomed by all save 'those whose business or ambition it was to improve human nature and who seized with delight such rich materials for what they were pleased to call the multi-racial society'.*

In this critical area at least, Britain now seemed well on the way to policy-making by opinion-poll roulette—with the wheel sometimes 'fixed'. Two months later, the Conservative leader who a little over a year earlier had disgustedly sacked Powell, was self-protectively demanding new legislation in which immigrants would be admitted 'for a specified job in a specified place at a specified time' and would have to have their permits renewed annually, and whenever they changed jobs.[19]

---

* In point of fact, in the first two months of operation, the Government's assisted repatriation scheme received from the whole of the United Kingdom twenty-five applications. (London *Times* 30 December 1971.)

In its application of new regulations, the Labour Government reflected hardly less the same bleak, restrictionist mood. Britain had come a long way from the proud 'citizens of the Commonwealth' concept so stoutly maintained less than ten years ago. For the first time in Britain in time of peace racial animosity and colour prejudice had emerged on a more or less open, nation-wide basis. In 1968, commented the Institute of Race Relations report, *Colour and Citizenship*, the 'liberal hour' ended. 'Honourable men shrug their shoulders, and turn aside ...'

Waiting, no doubt, for the next Gallup Poll.

The Forum and the Market

An eighteenth-century sage once observed: 'Let me write the people's songs, and I care not who makes the laws.' The projection by the opinion polls of the racial phantasmagoria of Enoch Powell in England—or of the 'Red' phantasmagoria of Joe McCarthy in America somewhat earlier—suggests that the time was fast approaching when this maxim might be reminted to read 'write the people's questions'.

Yet more important than the questions asked are the questions not asked. They are the questions whose answers wriggle under the statistician's pin, which resist being fitted even into the most 'sensitive' scalogram. They are the questions which won't put an edge on headlines to sustain the newspaper Numbers Game; which, indeed—outrage of outrages—perhaps won't chime with current 'news' or 'trends' at all. They are, writes Dr Gallup, 'typically ... questions which are of interest only to a small sector of the population ... on which the majority ... have not had a chance to formulate an opinion.'[20]

They are, in short, the questions which form the live elements, the yeast, in the movement of true opinion; they carry those elusive 'congenerics' without which ideas are merely words as whisky is merely yellowish alcohol; without which 'opinion' moves in narrowing circles from the plagiarism of the headlines to the plagiarism of the polls and back again.

In mid-1969 the Institute of Race Relations published the

results of a longer-term, deeper-going attitude survey, not anchored to Powell-inspired questions or the needs of the news headlines. It showed that even in towns with a relatively high proportion of coloured residents only one-tenth of the white public showed strong race prejudice, while 35 per cent showed no prejudice, and 38 per cent were 'tolerant-inclined' (as against 17 per cent 'prejudice-inclined').[21]

Inadequate as all such statistical exercises must be, the different approach does at least hint at complexities—in refreshing contrast to the unthinking hostility, blank rejection, and even paranoia given substance by questions differently phrased and selected. But by this time much damage had been done. The continuing, many-sided, wary processes of interchange, whether in workaday human situations or through various community agencies, were short-circuited by heavily publicised poll stereotypes, supported by massed statistics of 'opinion' much of which did not in reality begin to qualify for that honourable name.

Are we not, in short, in some danger of bringing into operation a sort of 'Gresham's Law of Opinion' in which the counterfeit, now minted in such vast quantities and with such professional verisimilitude, drives out the true coinage? In his *American Commonwealth* James Bryce points out that 'in a free country especially, ten men who care are a match for a hundred who do not.' It has been the achievement of the opinion polls to erase this distinction, equating the responsible and the irresponsible in their mechanical, semi-automated processes of opinion manufacture.

In 1962 the 'Port Huron Statement' of the radical student body, Students for a Democratic Society, spoke of America as a 'democracy without publics' in which politicians responded 'not to dialogue, but to pressure'.[22] They called for the 'democracy of individual participation', 'organised to encourage independence in men' and to permit 'that the individual share in those social decisions determining the quality and direction of his life'.

They called, in short, for the antithesis of the mechanised mass public opinion poll: for a deliberative interchange rather than an addition sum. An early model of that basic democratic process is the jury, still today a bastion of political liberty despite its modest size and relative obscurity. Like the respondents in an opinion poll, the jury is recruited by 'sampling', if of an unscientific sort, but unlike them its members are not expected, in isolation from each other, to offer an instant verdict. Equal at the vote, in discussion in the jury room some will have a greater contribution to make to the common stock than others, and will make it. Furthermore, in requiring near unanimity,* the jury system recognises that truth is not necessarily easily attainable or readily divisible into percentages, even with decimal points. Conspicuously made up of individuals, the jury yet transcends them and, at least in long or complex cases, may acquire the collective personality of a miniature society. Indeed Maitland in his *History of English Law* declares: 'The verdict of the jury is not the verdict of twelve men; it is the verdict of the community.' It is easily possible to feel that this could stand for a jury, but for an opinion poll it would be totally lacking in conviction.

No doubt such a body of ordinary citizens is not impressive to the eye. No doubt its methods are fallible. Certainly they are unscientific. But not even in America—where trial by jury is in fact enshrined in the Constitution—has there yet been any proposal for its replacement by national opinion polls, whatever the cases of Senator Edward Kennedy at Chappaquiddick and more recently of Lieutenant Calley of Mai-Lai fame might suggest.

And yet increasingly, in this way, we find moral—even legal —'verdicts' given through the anonymous, remote, mechanical process of the opinion poll. After the news of the Mai-Lai massacre the Louis Harris poll reported that only 22 per cent of Americans expressed moral repugnance at the idea that American soldiers may have gunned down innocent civilians. But we have no means of knowing what this very solid-looking, value-building 'fact' really implies. We do not know just what considerations were in the diverse minds, or how far, in the total abstraction of the polling situation, 'Americans' sampled really apprehended, or 'felt' the situation. But two Harvard scholars,

* When in Britain in 1967 it was decided to legalise 10:2 jury verdicts in criminal cases—to avoid the danger of a single jury member being 'got at' —the step was only taken with great reluctance amid misgivings expressed by eminent lawyers.

who commissioned a later poll,* have suggested that the poll verdict reflected a 'pathological' sense of powerlessness, a feeling, particularly among the less prosperous, of being 'pawns, not independent agents'.

It is a feeling opinion polls have been able to do little to correct, because a verdict thus reached resembles a price made in the market more than a judgment arrived at in a democratic assembly. And not a market of the small French town sort, where the housewife palpates the Camembert or cross-examines the stallholder, face to face, on the provenance of his aubergines, but the infinite, complex, elusive ticker-tape markets of Wall Street or Throgmorton Street. Like the Stock Market indices, opinion poll percentages are precarious balance points, statistically expressed, of innumerable causes and accidents, moods, fragments of information, hunches, rumours, wild and true, knowledge and experience, and the human cycle of hope, greed, fear and panic. The market moves in its mysterious way its wonders to perform. It is always right, even when it is manifestly wrong. A price—or an opinion—thus established is one of the sharpest of 'facts' and if it is arrived at by sufficiently complex and untraceable processes, the mid-twentieth-century mind will abase itself before it.

In this world of the Numbers Game, counterfeit opinion, just because it can be 'priced' with such precision, may enjoy a mechanical advantage, favouring the operation of Gresham's Law of Opinion. Figures support figures, existing, as they do, in their own right. As late as January 1954 when—to quote Joe McCarthy's biographer, Richard Rovere, 'the record was pretty well in, and the worst as well as the best was known,' the numerical professionalism of the Senator from Wisconsin (205, 121, 116, 81 or 57 Communists in the State Department) had made such an impression of 'objectivity' that the Gallup Poll indicated that 50 per cent of Americans still held a favourable view of him, and only 29 per cent definitely unfavourable.

To challenge 'facts' so weightily and repeatedly totalled as to seem almost facts of nature called for no little courage and

---

* The poll concluded that two-thirds of 'Americans' thought most of their fellow citizens, if so ordered, would 'shoot all the inhabitants of a Vietnamese village suspected of aiding the enemy, including old men, women and children.' More than five-eighths thought Lieut Calley should not have been brought to trial at all. (*Time* 10 January 1972).

resolution. But in March 1954 in his television programme, 'See It Now', Ed Murrow asked the questions which were not asked. In what a veteran critic duly deplored as a departure from that objectivity proper to an American reporter, Murrow put film-clips of McCarthy's speeches and inquisitions alongside his own queries and refutations. (The Senator was accorded equal time to reply.) Murrow concluded a historic broadcast, heard by 59 per cent of America's adult population, with the sombre words: 'There is no way for a citizen of this republic to abdicate his responsibilities.'

Wasn't there, though? It is indeed the characteristic of the Numbers Game both in the United States and in Europe that there have never been so many—eminently respectable—ways.

Nervous about the effect of Murrow's courage on the network, Dr Frank Stanton commissioned a follow-up Elmo Roper poll. It showed that one-third of those who had listened to the programme considered that McCarthy had proved Murrow a Communist or had raised doubts about him.[23]

And you can't argue with 'the figures,' can you?

6

UNTOUCHED BY HUMAN HAND

To the scientist, the unique individual is simply the point of
intersection of a number of quantitative variables.
> Professor H. J. Eysenck, *The Scientific Study
> of Personality*, London, 1952.

Perhaps statisticians themselves have not always fully recognised the limitations of their work ...
> Professor A. L. Bowley, *Elements of Statistics*,
> 1937 edition, first published in 1901.

... with the efflorescence of 'attitude research' ... nothing
was alien to statistical inquiry ... quality was relegated to the
defensive ministrations of humanists-in-retreat.
> Professor Daniel Lerner, *Quantity and Quality*
> New York, 1961.

In his introduction to Zeisel's well-known primer, *Say It With
Figures*, a leading American sociometrician, Professor Paul
Lazarsfeld, writes: 'Modern social life has become much too
complicated to be perceived by direct observation.' Of course
some of us in occasional moments of what used to be called
insight may still now and again feel that we perceive a thing
or two, but as educated persons we know that, however authentic they may appear, such glimpses are merely impressionistic,
tainted with the subjective, and even, very possibly, riddled with
value-judgments. We are no longer so naïve as to trust our out-
ward—much less our inner—eye; we realise that to penetrate
to social reality today requires a wide range of 'indicators',
mainly numerical, a sort of statistical radar screen monitoring
our lives, installed, operated, and interpreted by the task forces
of social science. If we want to know where we are, and where
we're headed, it is they whom we must ask.

A glance at almost any day's newspaper will confirm this.
Social psychologists working in Group Dynamics advise army
officers how to put together a good combat team ('it's better if

they like each other');[1] cost-benefit economists, emerging from months of exacting calculation, inform us whether we prefer, on the whole, being stifled by petrol fumes in traffic jams to being suffocated in commuter trains; and after rigorous experiment the Leicester University Mass Communications Centre reveals that television audiences absorb programmes selectively so as to reinforce their existing prejudices—the proverbial wisdom on this subject, although vividly stated, having, it seems, hitherto lacked statistical proof.

With the advent of television the inquiring—'layman'—reporter eliciting the inner truth of things from the appropriate social scientist—and there is almost always an appropriate social scientist—soon became as standard a form as consultation of the Delphic Oracle in Ancient Greece. In the United States it was impressive to note the confident way in which men as unacademic as Presidents Johnson or Nixon now turned to the nation's social scientists for the answers to its moral, social, and cultural—as well as its economic—problems. For President Johnson's National Commission on Law Enforcement there were 175 regular consultants, who guided five national surveys. In addition hundreds of social scientists were brought to Washington from all over the country to give evidence. Complete with statistical tables and charts, the final report, *Crime in a Free Society*, filled ten large volumes.

It appeared in February 1967, and was followed, less than eighteen months later, by the assassination of Robert Kennedy. On the very eve of that tragedy, President Johnson appeared on radio and television to announce yet another National Commission, this time 'on the Causes and Prevention of Violence'. 'Supported by suggestions and recommendations of criminologists, sociologists and psychologists,' the President told the nation, 'we may hope to learn why we inflict such sufferings on ourselves, and, I hope and pray, how to stop it.'

The First Amendment has always ruled out an Established Church in the United States. Yet the promptness and almost ritual manner in which the Head of State invited the counsel and ministrations of the incumbents of the country's social science establishments, often massively supported by the Foundations, suggests that this gap may now have been filled. The case of Britain is somewhat different. For not only does she

have an Established Church, but also a Mandarinate, in the
Administrative Class of the Civil Service, possessors all of 'well-
trained minds', formed by Oxbridge and the Classics, which it
had hitherto been assumed could take in their easy stride all the
problems the modern world had to offer. But now 'effortless
superiority' too was in retreat.* It was vanquished by the superiority of all-too-visible effort, as in the oncoming age of the
PhDs each new social science, specialism and sub-specialism,
staked out its territorial claim and paraded its array of authoritative statistical indicators, the keys to the kingdom.

If the *No Trespassers* signs were on the whole discreet and
the fences merely mildly electrified, the areas enclosed were
vast and central. Yet the pre-emption was finally accepted, even
welcomed. For absurd as it might have initially sounded the
term 'social scientist' had a long-run potency. It seemed to
promise the integrity, the objectivity, the accumulating achievement that had distinguished the natural and physical sciences.
Accepting the 'permissive' recommendations of the social
scientists who gave evidence before the American Commission
on Obscenity and Pornography, a Methodist clergyman on the
Commission, G. William Jones, wrote: 'Science has often
proven to be God's handmaiden in the quest for Truth.'[2]†

Sustaining 'handmaiden' status is no light task. It has meant
strenuous insistence on rigorous standards of proof, and, more
particularly, on the 'scientific' precision of number. The crux of
the matter is stated in the quotation carved on the face of the
Social Science Research building in the University of Chicago.
It is a dictum of the great physicist and mathematician, Lord
Kelvin—and what better warranty could one offer than that?
IF YOU CANNOT MEASURE, YOUR KNOWLEDGE IS MEAGER AND
UNSATISFACTORY.

No one reading the literature of social science today can
doubt that this maxim has been taken to heart. The emotion of
'anxiety' may appear one of many imponderable aspects of the
human condition. No matter; galvanic skin response (GSR) or
the Palmer Sweat Index will, objectively and precisely, provide.
Crime in our times may appear a phenomenon of baffling com-

* 'Effortless superiority' was the mental posture Dr Benjamin Jowett,
Master of Balliol College, recommended to his young men.
† For Commission, see next chapter.

plexity, an intricate and shifting web of social, economic, psychological, biological, political, legal, moral—and even spiritual—factors. But today it is statistics that, rightly or misleadingly, give it social definition, and, increasingly, diagnose, prescribe and predict it.

For as the eminent social statistician, Mr Leslie T. Wilkins puts it in *Prediction Methods in relation to Borstal Training*: 'Whilst it is likely that there are some unmeasurable factors [in delinquents' case-histories] it has yet to be shown that they are uncorrelated with measurable factors.'

This is the authentic note, the motto that would appear to be engraved on many a social scientist's heart. Unlocked, and now indeed computerised, Karl Pearson's 'treasure chest' of correlation which, as we have seen, had already yielded up the unexampled riches of Dr Kinsey, continued to pour forth its highly miscellaneous contents. Statistics and statistical techniques now took on a dynamic role over a wide area of the social sciences, determining their direction—and, therefore, increasingly, ours.

The resultant diagnoses, indices, indicators, made endless sharp headlines. But the side effects of this somewhat desperate preoccupation with measurement received little attention. In fact, they are insidious and continuous, and fall into two main classes. One class of side effects arises because what in the end is measured, as a result of the vast ingenuity and effort poured into the task, is often not quite what it is supposed to be, but something a bit different. A second source of confusion, compounding the first, is that factors which persist in defying the ingenuity of would-be measurers, which resist both the calipers and correlation, are quite likely to 'evaporate' and be effectively denied existence.

Such incidental distortions, taking place, unnoticed, in, so to speak, the statistical machine, may be multiplied a hundredfold by the current predilection of social scientists for prediction— for, again, the ability to predict accurately, even more than the ability to measure, underwrites scientific status. (When possible, it is also, of course, very useful.)

Whether the prediction is economic, political or social, it is again statistical science that provides the key instruments. Statistical correlation techniques, for instance, provide the basic mechanism of the system for predicting the onset and course of delinquency in human beings, pioneered in the United States by that remarkable husband and wife research team, Sheldon and Eleanor Glueck. Their life work is, indeed, an outstanding monument to the dynamic power and reach of statistical analysis. They began, almost fifty years ago, with a meticulous investigation of all the ex-inmates of the Concord Reformatory whose sentences had expired in the years 1921 and 1922. With the subsequent outcome for each of these cases known, they and their team worked back over the records and life histories of each of these 510 men—some of whom had now been re-convicted—visiting their homes, interviewing relatives, checking official reports against personal accounts, and all the time seeking to pick out the factors which seemed linked to successful rehabilitation or continuing failure.

Over fifty factors were selected. Each of these was then statistically correlated against the fact of failure (or its absence, success). From this mathematical sieve were then collected the factors which, overall, showed the highest positive correlation with failure; and from these, after various technical checks, the first crime or delinquency prediction tables were constructed.

Later, the Gluecks, verifying their predictions, followed up the same offenders for a further five years, then for another five, refining their prediction tables again. In yet another survey, this time of 500 delinquent women, they sifted no fewer than 285 factors to arrive at those few which correlation showed to be most predictive.

By the 1940s the Gluecks were moving on a stage further, from the prediction of the recurrence—or non-recurrence—of delinquency to the prediction of its emergence—in children entering school at the age of six or seven. The Gluecks claim that their social prediction table, in which numerical weights are given to the critical factors, supplemented by a Character Structure (Rorschach) table, and another based on character traits scored at a psychiatric interview, would enable the teacher at an early stage to distinguish betwen the 'true delinquent' (delinquent-to-be) and the 'pseudo-delinquent' (or the mere naughty boy).

The New York City Youth Board hailed this predictive instrument as a 'Geiger counter' by which potential juvenile crime could be nipped in the bud through preventive treatment. In the District of Columbia schools the Youth Council stated that a refinement of the Gluecks' social prediction table had proved 100 per cent accurate in predicting non-delinquency and 81 per cent accurate in predicting delinquency.[3]

This was hardly surprising, for what the Gluecks had mainly established was that what any commonsense person would call 'bad homes' will produce 'bad' children. Their analysis of the backgrounds of 500 delinquent boys and a control-group of 500 non-delinquent boys in Boston state schools showed that only half of the delinquents had been raised continuously by both parents. In one in three of their homes there was an open breach between the parents, as compared with one in seven in the control group. The delinquent homes scored high for 'plan-lessness', maternal neglect, paternal indifference, irresponsible budgeting and so on.

If such devoted and monumental mathematical exercises so often tell us something we have long already known 'in our bones', modern man, nevertheless, is now unable to act without the statistical evidence. Unfortunately, as we shall see, he may be even less able to act *with* it. He may find it an excellent substitute for action.

Hopes and (Statistical) Expectations

Recondite as such exercises may appear, they are in fact in the mainstream of the evolution of the science of social statistics. Man is bulked, and from the vast array of human digits statistical constancies are derived—as in the classic early instance of life expectation tables. When no more is involved than the price of annuities or insurance premiums the actuaries may fairly reign supreme; but in more complex social areas the relationship between the moralist and the humanist and the industrious statistician may be a less easy one—and may even raise, unbidden, the question of a future take-over bid.

In postwar Britain the logic of the Gluecks' prediction studies was pursued a stage further in the researches of Mr Leslie T. Wilkins working with the criminologist, Dr Hermann Mannheim. On the Glueck model they worked backwards over a large representative sample of ex-Borstal inmates from a point in time three years after the conclusion of their sentences. But Wilkins, a dedicated professional statistician, brought to the enterprise the sort of 'rigour' hardly found in the still humanistic Gluecks whose first 'prediction' book had at its heart a detailed, finely observed, and no doubt on occasion 'impressionistic' (and even 'anecdotal'—the ultimate pejorative) 'Sheef of Lives'. In the Wilkins–Mannheim research no field investigators dealt in such 'intangibles'—not to say unmeasurables—as the atmosphere of the delinquents' homes or the discipline of the father. Wilkins sternly confined himself to documentary sources, official records, and case reports, combing these for factors which could be objectively quantified. In their final calculation and prediction table 'work habits' are reduced to 'longest period in any one job'; relationship with parents becomes simply 'living with parents or not'. Actual human delinquents and case-histories are firmly relegated to a small-type appendix where their function is to confirm, or not confirm, the statistical operation.

This is purely a matter of statistical logic and mechanical calculation. From the records of the 720 sample cases sixty items capable of measurement are picked out. With the aid of punched cards the scores for these were correlated with the known outcome, and the highest relationships above chance selected as predictive. In the end, ten major factors stood out. But in the untidy way life is apt to have, they overlapped. With this statistical 'inefficiency' the Gluecks had been unable to cope. But now the overlapping factors were put into a 'correlation matrix', were further processed, and then redefined so that only the unique element of each remained.

Thus, after a vast amount of complex, mechanised calculation, fully verifiable only by another statistician, emerged the seven predictors of the Mannheim-Wilkins Borstal Lads Table, each item allotted its appropriate weighting, and collectively furnishing 'a tool of prediction which could be used with the same effect by anyone with a knowledge of simple arithmetic, and irrespective of his skill or experience in other fields.'[4]

With this final table completed, offenders could be placed, according to their statistical scores, in one of five sections, descending from 'A' at the top, where the chances of 'success' (i.e. freedom from recorded reconviction) were seven in eight to 'D' at the bottom where the chances of *failure* were likewise seven in eight. In between, in the middle area, lay an 'X' class, roughly a third of the total, where the statistical chances were considered too nearly equal to be predictable—yet.

It is claimed that such statistically-engineered prediction—untouched, so to speak, by human hand—has been shown to be more than twice as accurate as the forecasts made by Borstal masters and other immensely experienced assessors who actually knew the boys concerned well. Nor is this unusual. It has been repeatedly shown that this sort of applied statistical 'logic' can, even in complex human situations, often beat expert and experienced intuitive observers.*

In California at least the point was taken. In the late 1950s Leslie Wilkins was invited there to assist the Department of Corrections in installing an elaborate predictive system for its inmates, establishing what were called 'base expectancies'. 'With the aid of computers,' reported John P. Conrad in 1965, 'it will not be long before every living soul committed to the Department will be rated as to his expectancy of future major criminal activity.'[5]

Theoretically at least, such 'expectancies' should offer a yard-stick against which the progress of techniques of 'treatment' can be measured. In both British and American penal establishments 'expectancy' figures, attached to each case, should make it easier to treat like with like, underwriting earlier parole. In fact, although California has America's highest crime rate, only 13·5 per cent of its convicted offenders go to prison, and in the last two years the prison population has actually fallen.[6]

Theoretically again, such scientifically-established prediction tables and individual 'scores' should provide a compact guide to the judge in sentencing. But it is here that a certain dilemma declares itself.

It is, increasingly, the dilemma of our times.

* After examining about twenty separate studies in which statistical prediction and intuitive or clinical prediction could be compared, the American psychiatrist, P. E. Meehl, concluded that in about half the instances the clinician's performance was 'definitely inferior', while in the other half the two approaches gave about the same results—*Clinical versus Statistical Prediction*, 1954.

A judge, determining a convicted offender's fate, might find that he fell in the lowest class on the prediction table with only one chance in eight of 'going straight'. But what the table omits to tell him is whether this single future 'success' is the man awaiting sentence. It could be: after all, one-in-eight of all the prisoners found guilty in a year would amount to a sizeable slice of humanity.

Certainly, the judge might then take the offender's actual score, from the addition of the various factor-scores, and read off on a graph the exact percentage chance. The precision is impressive, indeed it is dangerously so—because it still does not relate to a real *person*. Once again we are looking upon that very strange, very contemporary entity, a statistical man, end-product of highly sophisticated calculations, fully intelligible only to mathematicians, and based on a high degree of abstraction.

And it hardly ends there, with just the occasional, 'inevitable' error of judgment. Erich Fromm, the psychologist, has divided humanity into those who are 'I am' and those who are 'the sum of other people's expectations'. When that 'sum' emerges from a computer, with all the authority of that Mystery, and arrayed in the immaculate white coat of Science, 'expectation' acquires a peculiar force which may overwhelm and devour many an 'I am'.

As we have already seen in a number of instances and will see in more, it is the statistical artefacts which may then become the 'I ams', moulding both individuals and society in their patterns.

For today, after the statistical shuttering has been removed such statistical artefacts are free-standing, existing in their own right, like works of architecture or art, but more insistent and influential.

It is all the more eerie—and all the more insidiously effective —in that there is about the whole thing a substantial element of neo-Occultism, or sleight of hand. The calculations of factors in the Wilkins-Mannheim prediction tables do not necessarily point to causes of crime and recidivism. On the contrary, as Wilkins himself points out: 'A large number of correlations we used ... may be spurious correlations or correlations which exist only because there are other factors associated both with our criteria and the factors concerned.'[4] In short, for all we can actually *see*, statistical logic stabs the finger of 'expectation' at the 'predictee' as in a scientific version of the African witch doctor in action.

All in all, it is perhaps hardly surprising that prediction tables have been slow to come into regular use on the judicial bench. 'Nevertheless,' wrote John P. Conrad, reporting on his researches to the California Department of Corrections, 'predictions cannot be longer resisted in an electronic age. It is up to the correctional administrator to make sure that the place of the computer is consistent with its capabilities.'[5]

The note is familiar, almost ritual: one begins indeed to wonder whether some such sentiment may not provide the Famous Last Words of our civilisation. Wilkins and Mannheim also echo the Gluecks in suggesting that their prediction scoring tables should be employed 'as a servant rather than a master', a phrase, the reader may recall, once used of Audience Ratings at the BBC. Wilkins indeed goes further and suggests that it is permissible for an assessor to take into account 'factors which were not included in the [prediction] equations, or examined in their construction.'[4] But it cannot be easy to know what *was* taken into account in such wide-ranging factors as 'Living in an Industrial Area, *add* 8 points.'*

For is there not in all this a deeper, central—and today a symbolic—incompatibility? As a British criminologist, Dr Howard Jones, rightly points out: ' "Prediction tables" are based on past experience and can be expected to work only if applied strictly . . . Insurance companies use similar methods in fixing their premium rates.'[7] To bring unvalidated 'impressionism', mere human hunches, however strong, to the adjustment of advanced statistical mathematics where the degrees of probability have already been precisely calculated appears an unforgivable presumption.

'Only connect . . .' advised the humanist.

'Only correlate,' cries the busy statistician.

---

* The Third International Congress on Criminology, held in London in 1955, concluded that to place prediction tables in the hands of judges might be dangerous since they might be tempted to rely on them. Rather should they be used as an aid to professional workers reporting to the court after studying the case.

It is not quite the same thing. The symbolic blinding of the figure of Justice may yet take on a new—and somewhat more literal—meaning.

Expanding Empire

Many of the builders of this contemporary empire of number seem to be driven by a sort of statistical imperialism, a sense of mission which Charles Dickens detected as long ago as the 1850s when in *Hard Times* he satirised these 'new men', the statisticians, 'ready to weigh and measure any parcel of human nature, and tell you exactly what it comes to'.

It is notable that throughout the Mannheim-Wilkins British Home Office Report on prediction methods such statistical procedures are characterised as 'rational'—in contrast, it seems to be implied, to all the less tidy, less mathematically watertight approaches. Admittedly, statistical analysis still leaves unsolved problems. But closing these gaps, it seems to be implied, is only a matter of time. If, for instance, the chances of the middle third of reformatory inmates appear too nearly equal for prediction, this, we are told, can probably be taken care of by selecting a new range of factors and a new set of correlations. If the scoring scale validated by statistically processing one slab of delinquent humanity loses accuracy with time and later intakes, this can be adjusted by sampling techniques similar to those devised for quality control in industry. One recalls, with some trepidation, Dr Gallup's memorable phrase, 'the continuous audit'.

Confidence in the omnicompetence of statistical reasoning grows by what it feeds on. In a world so deplorably full of imprecision, so riddled with the subjective, there can be no lack of new fields to conquer. Moving from the prediction of delinquency to the somewhat more difficult area of its treatment, there is for instance all the uncertainty surrounding the business of probation.

'Probation under one probation officer,' explains Mr Wilkins 'is likely to be different from probation under another. No matter how intensive the training the officers receive, they are different personalities ... At present, it is generally believed to be personal contact between officer and offender that has a therapeutic effect, but very little is known of what elements make up this or any other treatment system.'[8]

Although he would probably doubt whether it could ever be possible to analyse on paper the complex chemistry of such a relationship, a layman might feel that the best way of getting a shrewd idea of what lay behind a probation officer's success, or lack of it, would be to allow an experienced and sensitive outside observer to watch him on the job. It turns out, however, that this would be naïve. 'Impressionistic concepts cannot be used in scientific analysis. Nor are they really effective for communication because they have a tendency to change meaning with context and with the personality of the user.'

No, the answer must be to apply to the relationships between the probation officers and their clients the methods of objective measurement and statistical analysis. The project has, in fact, been under way in England for more than ten years now. In a pilot study in Middlesex 'treatments' were divided between eight permutations of 'Control' and 'Support', either of which might be 'Situational' or 'Individual', and might be rated 'High' or 'Low'. Probation Officers themselves were quantified by age, length of service, length of training, and prevalance of certain attitudes—all extracted from their replies to questionnaires. Offenders were likewise scored, together with a scientific measure of their environment, including a Stress Score, to be correlated with the California Jesness Inventory of 155 personality and 'adjustment' items.

Finally, the machinery of correlation could be put to work, although, as usual, it brought its problems. 'Success rates tended to be associated with Low Control, but many of these cases may have been good risks who did not need much control in any case.'[9]

In the meantime tape-recordings of talks between probation officers and offenders were being analysed by 'Resource Process Analysis' in an attempt to show 'the relative occurrence of different kinds of interaction, and to develop indexes of successful control and successful interpretation'.[10]

In the Mannheim-Wilkins report on Prediction it is suggested at one point that the reason reformatory masters, probation officers and so on compare so poorly with statistical prediction in the accuracy of their prognoses may be that the 'good therapist' owes his success to being an optimist. He is the sort of man in whom hope obstinately goes on triumphing over statistical expectation. This may very well be. But one can perhaps be pardoned for wondering how long such necessary and creative hope will survive the steely scrutiny of mechanical correlation and the continuous audit. Retiring in 1972 after 35 years in the probation service at a central London magistrates' court, chief probation officer, Mr Charles Morgan, said: 'There is a very real reason for worrying now whether we are meant to be more interested in paper or people.'

A New Dualism

Increasingly, we find ourselves caught up in a new contemporary dualism; there is the muddling-on, verbalising, impressionistic, human old world down here, and there is that Other, Finer, Rational World to which the better statisticians have already been called. Communications between the two can be tenuous.

For instance, in his *vade-mecum*, *Say It With Figures*, Professor Hans Zeisel warns against the onset of that treacherous moment when a finely constructed statistical index is to be translated into mere words—given a name 'which in a more or less vague way purports to define the object'. 'Extreme caution,' he points out, will be needed, because 'the language of words is much less accurate than the language of numbers.'[11]

If, for instance, one wrote that there had been a 5 per cent advance in a standard-of-living index, the meaning would be immediately transparent, but if one spoke of 'happiness' a deplorable ambiguity, which apparently escaped the attention of the authors of the Declaration of Independence, would descend. So much so that a few years ago the National Opinion Research Centre of the University of Chicago felt obliged to design a survey to clear the thing up.

'... the underlying assumption of this research,' its report explained, 'is that there is a dimension called variously mental health, subjective adjustment, happiness, or psychological well-being ... At present there is neither a generally agreed name for this dimension, nor agreement as to the appropriate methods of deciding where a particular individual should be placed on such a dimension ...'

A pilot study was therefore mounted covering 2,006 persons, of whom 24 per cent professed to be 'very happy', 59 per cent to be 'pretty happy' and 17 per cent to be 'not too happy'. The vagueness of words compared with figures was immediately apparent, and the endeavour to fix and calibrate the 'dimension' was pushed ahead.

Among other things subjects were asked, for instance, 'Did you feel "on top of the world" at any time last week?' a question of bracing precision, permitting the answers 'once', 'twice', and so on, although whether this feeling arose from achievement, love, whisky, self-conceit or LSD does not appear to have been taken at any point into the total. Instead, the correlation mill was put to work, and various numerical truths emerged. It was found, for instance, that when income is the same, the better educated are less happy than their less cultivated counterparts; that single men are twice as 'unhappy' as married men; and that 'only 4 per cent of those under 50 as opposed to 21 per cent over 50 worry about growing old.'

And the conclusion of the pilot study—expressed, of course, with the 'extreme caution' requisite when descending from the higher realm of numbers to the uncertain world of words: 'Happiness is the result of the relative strength of positive and negative feelings rather than the absolute amount of one or another.'[12]

One may fear that if Maurice Maeterlinck's Blue Bird of Happiness were released today the small creature would decline to take wing, and would presently be found on the floor of its statistical cage, rigid with 'rigour'.

VACUUM-PACKED

Figures are as flexible as words and more deceptive because they carry with them the air of neutral facts...

One of the most persistent factors in holding back science has come from the very success of science itself. It is the belief in science as a means of attaining to absolute and permanent knowledge.

J. D. Bernal, *Science in History, Volume 4*:
'The Social Sciences'.

It is a curious circumstance that the more often the social sciences fail in their self-imposed task of measuring the unmeasurable in order to prove the unprovable, the greater becomes our dependence upon them, the more total the abdication of our own capacity—and responsibility—for judgment. 'Commonsense', which really refers to insights derived from the collective experience, takes flight, or becomes a dirty word.

If this appears exaggerated let us consider the case of television—from which the children of the nation today gain some of their earliest experience and values. The character and quality of such a socialising medium is clearly a matter of inalienable public concern; a matter on which every citizen has the right, the capacity, and indeed a responsibility to form a judgment. For years anxiety has been expressed about the number of beatings, knifings, kickings and other acts of brutality seen on the living-room screens, a phenomenon not unconnected with the well established fact that in conditions of unremitting competition the stepping up of the violence level affords the simplest method of pepping up a flagging audience rating.

Concern has naturally been increased because violence on the small screen has been accompanied by mounting violence in the streets. Yet by the contemporary occlusion in the body social and politic, it has become impermissible, most notably for all who wish to claim the status of 'well-informed men', to condemn an excess of media violence on simple common-sensical, aesthetic, or moral grounds. Although the 'effects' are expe i nced by us, it is understood that we must not testify to them unless they have first been 'scientifically' proved, and above all, measured. We are now under the bizarre necessity of being 'objects' before we can be persons.

Thus, in April 1970, w find the television-and-violence issue being debated for the nth ime in the British House of Commons —and once again left in the air in the now characteristic 'waiting for Science' (or Godot?) manner. A Labour Minister, Mr George Strauss, demanded to know whether it was not 'generally accepted that it is highly undesirable that children and adolescents should get the impression that acts of violence are normal and acceptable in civilised society?' The Home Secretary, Mr James Callaghan, agreed, but added: 'This raises difficult issues. For every piece of evidence on one side, I find an equal and opposite piece of evidence on the other side. What is clear is that there is far too little known about the consequences of the portrayal of violence on television ...'

Mr Norman St John Stevas, a member normally considered an exponent of the Roman Catholic point of view on moral issues, now asked for a programme of research—'so that the question can finally be decided.' The Home Secretary agreed to more research, then added, in a faintly puzzled way: 'I think it is regrettable that after all these years not more is known about the consequences of all this, although there has been a substantial programme of research so far ...'

Indeed there had. No medium in history had ever been subjected to so much continuous, social-scientific scrutiny. As early as 1964 a UNESCO descriptive bibliography of 'significant' research studies listed 165 reports, including a number of major national surveys.[1] We know, for instance, that as a result of the advent of television, children went to bed on the average 20 minutes later in Britain, 17 minutes later in Japan, and 11 minutes later in Canada. In Britain too the very birth of television was monitored in a monumental before-and-after, and with-and-without, survey, financed by the Nuffield Foundation, and directed by Dr Hilde Himmelweit. Children in five cities were investigated over four years, and so much data accumulated that forty 80-column Hollerith cards were required for each child.[2] In the United States the Senate Sub-Committee on Juvenile Delinquency took evidence from social scientists and reported on the effects of television in 1954, 1961 and 1964. In 1963 in Britain the Home Secretary established a Television Research Committee, and, three years later, its sociologist secretary, James Halloran, became also director of a new Centre for Mass Communications Research at Leicester University.

Yet, as the Home Secretary observed, despite the increasing sophistication of the methodology, the results of all this seemed strangely inconclusive. About all that seemed to emerge was that the normal healthy child was likely to survive even television, although the maladjustment of the already maladjusted and 'disadvantaged' *might* be deepened.* Each report duly concluded with the ever hopeful incantation: 'Further research is needed.'

There is no sentiment in the modern world with which it is safer or more respectable to agree. Few phrases confer a greater reputation for enlightenment at less expense.

'Further Research is Needed'

'Up to a point,' comments James Halloran, writing of the successive TV investigations in Britain, 'each wave of concern has been accompanied by its parallel inquiry.'[3] The correlation is, indeed, as they say, 'significant', but in what way? Is it the 'concern' alone that gives rise to the research—or could it also be that the research, in its labyrinthine progression to inconclusiveness, deepens and prolongs the concern, because it inhibits the normal judgments we all apply in life—and which indeed qualify us as being civilised and alive?

Perhaps the process is best seen in parable form. If a man eats too much and gets bellyache, his remedy is simple—to eat less next time. If, however, passing by this lesson of experience, he goes to the doctor with an elaborate list of sensations and symptoms (and it could well be elaborate), and the doctor, a

---

* This somewhat 'commonsense' conclusion seems once again to have been the main burden of the report of the US Surgeon-General's Scientific Advisory Committee after its two-year investigation of the effects of television violence, concluded in 1972. (*Time*, 24 January 1972).

young devotee of scientific medicine, embarks on a long series of tests, barium meals and X-rays, analyses of stomach juices, and so on, all the results of which are inconclusive, the patient's anxiety will certainly increase. If he is then passed on to another doctor, and the process repeated; and then to a third, and so on, with similar results, an incipient hypochondria may appear in the least morbid of subjects.

The difference between the doctors' diagnoses of a stomach upset and the social scientists of TV violence is that the doctors would, in part, use their own 'subjective' or clinical judgment, and, beyond that, would have truly objective or scientific tests to fall back upon, e.g. the chemical composition of the stomach juices in a test-meal. By comparison, although he often adopts a more 'rigorously objective' stance, the social scientist's apparatus for the establishment of the scientifically valid proof he insists upon is meagre.

The diagnosis of our TV confusions—and many others besides—should in fact swing wider—to include the social 'doctors' themselves. Sir Solly (now Lord) Zuckerman, then Chief Scientific Adviser to the British Government, pointed towards it in a recent book when he wrote: 'With so much science in the air, areas of interest or discussion become treated as scientific whether or not they are subject to the real discipline of the scientific method. We live in an age of more and more science, and also, alas, more and more pseudo-science. This is one of the unfortunate facts of our time ... The public is led to believe that anything is science that has numbers in it, or demands slide-rule rules, or is carried out by people with a PhD ...'[4]

In effect this insistence on quantification has narrowly limited the basic approaches open to social research. It has tended to mean either statistically engineered surveys, gathering material by questionnaires or structured interviews, or 'laboratory' simulation by human 'subjects' of real life situations. Either way, control groups and correlation and its extensions will probably be used to furnish 'scientifically valid' proof.

Unfortunately when applied by human researchers to human subjects, inevitably in social situations, these methods are as full of holes as an Emmenthal cheese.

As already suggested, however skilfully questions seeking to define human attitudes are designed, the variables are so many and so volatile and the possibilities of misunderstanding so rich that the quantifying investigator is visibly impaled on the dilemma that words are hopelessly vague, and figures—into which they are somehow to be transposed—are impossibly precise. The choice seems to lie between a confession of failure and a confidence trick, both glossed over with the invaluable phrase: 'further research is needed.'

In a recent British survey-type investigation of 'Television and Delinquency', directed by James Halloran, under the auspices of the British Television Research Committee, the delinquent group and the control group were asked, among other things, to give a single word to describe the sort of television programme they would like to see more often. 'Some interviewers,' we read, 'had considerable difficulty with this and similar questions'.[3] As well they might. This, however, did not prevent the answers—when they could finally be elicited—being categorised under such heads as 'Exciting', 'Thrilling', and 'Adventurous', leading to a tentative theory developed around the fact that 42 per cent of the delinquents wanted more 'Exciting' programmes as compared with 15 per cent of the control group; while 16 per cent of the lower-middle-class sample regarded 'Relaxed' as being the feeling a TV programme ought to produce in them, compared with a mere 4 per cent among the delinquents.

One will be told, of course, that such answers and results bear out other answers and correlations in the grand design, but as John Madge observes in his book, *The Tools of Social Science*, 'internal consistency is not quite the same thing as truth.' It is not indeed, but the mills of correlation grind rapidly, and with magnificent ambiguity, and, with their aid, it is possible to run up many an impressive multi-storey edifice which will carry the warranty of 'Science'.

The ambiguity, however, remains, even if its trailing clouds seem to give the edifice an extra touch of authenticity and splendour. For all the precision of the correlation coefficients, for every question they purport to answer they seem to raise three more. A positive correlation is established between 'heavy TV viewing' and delinquency. But ... is such a viewer a heavy viewer because he is a delinquent—or a delinquent because he is a heavy viewer? Or could heavy viewing be in part a substitute for the greater delinquency which would ensue without it?

Or does the explanation of the correlation lie somewhere else entirely? 'Further research is needed ...' To the layman who may study these heroic enterprises—and of course few really do—it may seem that here is a formula for perpetual regression; that today it is the social scientist rather than the philosopher who enacts the proverbial role of the 'blind man in the dark room' in everlasting pursuit of the 'black cat that isn't there'. And the thought may not be wholly confined to laymen. The late Professor T. S. Simey, the biographer of Charles Booth, characterised 'the fashion, if not cult' of multiple correlation as 'perhaps the most misleading technique ever relied upon by social scientists. Many people have gone far up blind alleys in this way and resolutely remained there, believing there is something innately "scientific" about work of this kind.'[5]

And when, in pursuit of the black cat of definitive truth, more refined techniques of statistical analysis, factor analysis, and so forth, are deployed, the researcher is more and more distanced from the subject of his pursuit,* and the real human world in which it exists. He rises as by a sort of mathematical levitation into that other, finer sphere, where black cats are clawless, mewless and abstract and human intelligence in all its richness and variety is a Quotient.†

Of Mice and Men

The natural scientist, conducting experiments with mice in his laboratory, enjoys a great many advantages denied the social scientist investigating human beings in his. Bred to uni-

---

* Even in Dr Himmelweit's classic, sensible, and even readable, major 1958 survey *Television and the Child*, although several thousand children are processed, only a handful were actually observed and chatted with by the investigators and these 'impressionistic' goings-on are swiftly passed over —like the teetotaller's sip of medicinal brandy in favour of the mass of children's 'objective'—because countable—answers to pre-set questionnaires.

† See Chapter 11

form bacteria-free strains in sub-micron filtered atmosphere, the natural scientist's laboratory mice are in effect living test-tubes. The setting up of control groups presents small difficulty, and a gestation period of 19 days and a life-cycle in which one year equals 30 in a man makes the replication of experiments reasonably speedy.

But when, as in the name of science he often must, the social scientist sets up human control groups matched to the groups on which his survey or experiment centres, the array of variables—the 'other things' which should be 'equal'—is formidable, their 'mix' is varying and complex, and, if for no other reason than that the degree of complication would over-whelm even a computerised operator, the number that *are* in fact matched is narrowly limited.

In the Television and Delinquency survey in Britain delin-quent groups and control groups were matched for the usual four factors—age, sex, IQ level, and class. But although both delinquents and controls were 'working class', this turned out to be a term which conceals critical differences; it was later discovered that the delinquents belonged mainly to the un-skilled or even 'rough' working class, a distinction which Charles Booth—who was not afraid of 'subjective' first-hand observation—had underlined two generations earlier. Nor, as the survey's director, James Halloran, himself points out, were the groups matched for the factor which appears most linked with delinquency—the 'broken' or 'incomplete' home.

But then as Mr Halloran goes on to observe 'an ideal [match-ing control] design may be impossible to achieve, and can indeed be sociologically nonsensical.' In fact by insisting on perfect matching one might well find oneself reduced to the single unique individual—and be left with no alternative but to study him or her sympathetically, in close-up, and at length. And this clearly would yield no scientifically acceptable 'proof', however illuminating it might turn out to be.

In the respectable category of work 'with numbers in it', there remains the direct 'laboratory' experiment, simulating the elements of real life situations. The general effort is to reproduce the traditional experimental patterns of the natural scientist. Human subjects in controlled conditions are exposed to measured or representative doses of the experience under investigation, and, in some way, the 'before' and 'after' states are calibrated, and measured. The results are supposed to be 'verifiable by replication'.

In one much-quoted experiment Seymour Feshbach, of the University of Colorado, was held to have demonstrated that television violence has cathartic value. Using students as his laboratory mice, Feshbach arranged to have one randomly selected group subjected to unwarranted insults, while the control group was treated in a 'neutral' manner. Half of each group was then shown a brutal prize-fight film, and half of each group a 'neutral' film. Later, 'word association and questionnaire measures' showed that the 'insulted' group who had seen the fight film were 'significantly less aggressive' than the insulted group who had witnessed the neutral film.[6] From this it was inferred that the 'fight' group had discharged its 'aggression' vicariously.

Unfortunately, the symmetry of this demonstration seems to have been somewhat disturbed by a broadly similar, also much-quoted, experiment by Albert Bandura of Stanford University, using small children. In this a group of children were introduced into a room where, while playing with the toys provided, they saw an adult beating up a 3-foot doll with a mallet, hurling it into the air, and verbally abusing it. Another group saw the big doll being beaten up on film. There were, of course, the usual control groups.

All the groups were then deliberately irritated, and, after a ten-minute interval, introduced into a room containing, among other things, a 3-foot doll and a mallet. The psychologists watched through a two-way mirror. The children who had seen the adult, whether in the flesh or on film, attacking the doll, took up their mallets and attacked the dolls. The controls hardly ever did so. The degrees of aggression were rated. 'The results,' concluded Mr Bandura, 'leave little doubt that exposure to violence heightens aggressive tendencies in children ... a person displaying violence in a film is as influential as one displaying violence in real life.'[6]

But is it not evident that such elaborately staged laboratory simulations are no less remote from the complexities of real life in society than the alternative approach of the statistical survey? A doll is not a person; television is not watched *in vacuo*; most seriously of all, human subjects, unlike Professor Skinner's pigeons in their boxes, are highly sensitive to all manner of extraneous suggestions, arising unspoken from the experiment or the experimenters themselves. It has been demonstrated in medical research that the administration of inert placebos commonly results in 25 per cent or more of patients claiming relief from physical symptoms in many common conditions.[7] How much more likely, then, is the power of unspoken suggestion to operate when the effects expected are wholly in the emotional or psychological sphere. Indeed Robert Rosenthal has shown that even with careful, trained researchers, and even though it may be impossible to see just *how* suggestions are actually conveyed, 'human beings can engage in highly effective and influential unintended communications—even under controlled laboratory conditions.'[8]

Of course, the social scientists engaged in such 'behavioural type' experiments take such difficulties as a challenge. Further research—or a new technique of control—is needed. Elaboration grows. In further experiments with student guinea-pigs Leonard Berkowitz of the University of Wisconsin introduced the idea of 'justified' and 'unjustified' violence. He allowed his 'insulted' groups to 'punish' their insulters (after seeing the 'violent' film) by depressing a telegraph key which administered electric shocks. This permitted measurement not only of the *number* of shocks given, but of the *duration* of each: two separate columns in the Volume of Aggression graph. Conclusion: 'It was clear that the people who saw the *justified* movie violence had not discharged their anger through vicarious participation, but instead felt freer to attack the tormenter.'[9]

But perhaps the most glaring defect of this 'natural science' approach which, via behavioural psychology, has now gained so dominating a position, is' that in its nature it must concentrate on the precise measurement of short-term 'effects', whereas what really matters and should concern us in such areas as television violence or pervasive pornography is long-term, cumulative effects, working on and through society as well as on our own characters and outlook.

And this is enough even to give a dedicated social scientist pause. As Dr Joseph Klapper, director of CBS Social Research, has put it: 'In the course of ten or fifteen years there are millions of influences on the individual—media influences and millions of other influences—and to determine the actual effects of each, and what effects would not have occurred if a given influence was absent—all this involves a research undertaking which is massive, and in certain ways perhaps impossible.'[10]

It would be foolhardy to place too much reliance on that reluctant 'perhaps'. For in an age when ministers of religion defer to 'scientific' testimony on the therapeutic qualities of industrialised pornography, the Archbishops and Convocation of the church of social science must pronounce or abdicate. And no disposition to do the latter has yet been detected: the flock, after all, cannot be abandoned.

Creating a Value Vacuum ...

'Today,' writes a New York schoolteacher in a book on television's influence, 'we are moving away from value-judgments, glib opinions, and the categorising of children's use of television in such terms as "good" or "bad".'[11]

The coupling of 'value-judgments' and 'glib opinions' is an interesting comment on the distance we have travelled. In fact, of course, if we are to retain control of our destinies, value-judgments are daily, and even hourly, necessary; they are of the very fabric of civilised life, and they do not, except with glib people, imply glibness, but rather all those delicate distinctions, stemming from a lifetime of experience and growth, which an individual human being makes in distinguishing, testing, and assessing a vast and shifting array of 'variables' and 'constants' such as would defy even the most 'rational' statistical process or the most rapid of computers. As Barbara Wootton, a British sociologist with long service as a magistrate, has well said: 'Individual conduct emerges from the richness of individual experience, and even among those whose lives by any standard would be considered poorest, this is rich indeed.'[12]

Nor probably is it altogether without significance that perhaps the greatest and earliest impact of the issue of violence in the American media (including comic books) stemmed not from laboratory simulations, but from the years of direct intimate experience of a clinical psychiatrist in New York child guidance clinics, years during which Dr Fredric Wertham says he has seen young children taking to crimes of violence in a manner never known before.

'Nothing,' Dr Wertham insists, 'can replace concrete clinical analysis of actual significant cases ... The super-scientific attitude which tries to reduce everything to the quantitative level ... leaves out what is truly human in the child,' and, after that, 'no statistical refinement can overcome the errors and ambiguities contained in the original data.'[13] To study a child, he says, you study a child—the whole child, background, satisfactions, and dissatisfactions, emotional life over a period, with follow-ups later.[14] You need time to gain a child's confidence, and subjecting a child to a set of questions is 'almost the opposite of clinical examination.'

'Nothing,' he says, by way of comment on the demonstrations of instant 'effects' school, 'has a longer preparation than an impulsive violent act.' After examining 300 cases in his clinics, he claims that children's 'inborn capacity for sympathy' which needs to be nurtured to flower, is being killed by television and other media which are inculcating 'not the pursuit of happiness, but the happiness of pursuit.'

Such views may be right, or they may be wrong. We are unlikely to be able to *prove* it according to 'rigorous' scientific standards either way. And unfortunately, the venerable injunction, 'Judge not that ye be not judged', has taken on a new and vastly extended significance. In an age when 'social life is too complicated to be perceived by direct observation', phrases like 'ringing true', being 'morally convinced that', 'feeling it in one's bones' have to be prefaced by apologies for naïvety.

But if, abandoning all such presumption (to 'feeling in one's bones' etc.) one seeks instead always the authoritative scientifically established verdict, when one has penetrated behind the precision of the decimal points, the long-promised proofs—'the Facts'—'all the evidence'—turn out to be remarkably elusive or insubstantial.

The net effect has been an ever-lengthening suspension of

judgment, a non-commitment and a progressive inertia rendered acceptable and even virtuous by those two ringing phrases of our time—'There is no conclusive evidence that ...' and 'Further research is needed ...' the Insh'Allah (God willing) of the dynamic, masterful, technological West.

... and Filling a Vacuum

But no less than Nature, society abhors a vacuum. If under the great bell-jar of the statisticians' normal curve, old values created and nurtured by individual and collective experience were now increasingly exhausted, others, more hastily derived from every-ready commercial—and other—self-interests, were now drawn in to fill the void. Admittedly these might often be more properly described as 'prices', but in a world in which the neutral colouration of the Numbers Game so predominated, this was a distinction which was less and less visible.

The process was illustrated with the sharpness—and the grossness—of caricature in the evolution of the great 'Is Pornography Harmful?' question which, together with the attendant 'liberalising' reforms, so strangely engaged the attention of the United States, Britain, and much of the Protestant 'Nordic' area of Western Europe at the outset of the seventies.

In retrospect, the American Commission on Obscenity and Pornography which reported in 1970 may be seen to have marked some sort of watershed in the heroic unremitting efforts of social science, in the nine decades since Galton, to prove the unprovable and measure the manifestly unmeasurable. At least one may hope that the Commission's monumental $2m report, with its extraordinary—yet classical—mixture of the modestly useful, the portentously obvious, the banal, and the utterly misleading, may turn out to have been the final territorial conquest of statistical imperialism which reveals the shakiness of the whole far-flung empire's foundations.

The fact that such a question as 'Is pornography harmful?' should be referred to a commission dominated by social science specialists, and monopolised by their methodologies, is in itself striking evidence of that empire's zenith. For the area thus pre-empted is clearly one that goes to the heart of civilised human life, embracing its whole texture and quality, both for the individual and the community. We know from common experience that there are some things which are life-enhancing and some which are life-impoverishing, and there really is little basic disagreement about what these things are; for this is something we feel in our inmost being, which answers to our specifically *human* needs. Unless we are conditioned or corrupted we know that hard-core pornography, systematic and in quantity, is either mad and meaningless, or demeaning and dispiriting, the more so because of its joyless and evil deliberation.

If, as is true, there are degrees in these things, they are still not statistical degrees; they are degrees such as may be experienced in a work of art. An experience may leave, as we say, various kinds of 'bad taste in the mouth' which will be as real as if they were physical. But it will be fruitless to analyse chemically for acidity. Nevertheless, to fulfil the expectations reposed in it, measurement 'science' must have. What, then, shall we measure to establish 'harm' or its absence? Dr Kinsey and modern technology supply the answer. Porn input, sexual output.* If we cannot take the temperature of the human spirit, we can meter the tumescence of the human penis. If we cannot calibrate self-fulfilment or true fullness of life we can collate totals of rapes and other sex offences and correlate them with twenty different specifications of pornography, verbal, photographic, and filmed. We can measure human well-being in terms of pathological degrees.

As so often, the ingenuity, dedication and organising energy displayed is extraordinarily impressive, and serves to make the irrelevance more tragic than comic. In one of the 83 constituent research studies the differences between the time taken by subjects viewing the same piece of pornography (a) in an audience (b) in private, were measured and graphed. ['Viewing time as a Function of Pornographic Rating.']. Eighteen different porno 'poses' were rated on a six-point scale for 'arousal effect': photographs' 'obscenity ratings' correlated with occupational

---

* Breathtakingly, the Commission observes at the outset: 'It was impossible during the brief life of the Commission to obtain significant data on the effects of exposure to pornography on nonsexual moral attitudes.' But even sexual attitudes are in fact defined in very narrow, materialistic terms.

groups; married couples exposed to a course of erotic films and asked to report on changes in their own and their spouse's sex patterns. And so on, and so on.

Then came the famous study to test the alleged 'satiation effect' bearing on the hopeful notion that after some time pornography lost its power to offend or excite and presumably sank into the general background of nature like the trees and the clouds and the sky. For fifteen days 23 student volunteers each spent 1½ hours a day in an 'isolation booth' viewing a collection of pornography (and optional non-pornography including the *Reader's Digest*) and wearing a loose robe to facilitate hooking up for electronic monitoring of 'penile volume and heart rate' (and subsequently of 'urinary phosphatase level') while being observed the while through a one-way mirror. The conclusion: 'subjects' interest showed a steady and significant decrease over the three weeks ... the time spent with erotica dropped to approximately 30 per cent in the eleventh session ... Nine weeks after the daily exposure sessions ended, all subjects reported boredom ...'[15]

Comment is surely superfluous. Few writers even of farce would dare so total a *reductio ad absurdum* in which all the deficiencies of the 'human experiment'—artificiality, self-consciousness, expectancy-effect, truncated perspective*—are so mercilessly exposed. Above all, perhaps, the deadly—blinkered and blinkering—humourlessness.

Our forefathers greeted the resurgence of life in spring by dancing around the maypole; the women of India anoint the stone lingams in their temples, the symbol of Shiva, with ghee and honey in mute celebration of the regenerative power of the universe; it has been left to the age of the Numbers Game to offer its latter-day Kinseyian phallicism in the form of a few descending lines on graph paper, improbably headed 'Satiation of Interest in Sexual Material'.

As ever, the most conspicuous feature of the Report is the sheer overwhelming weight and volume of the statistical 'evidence'—and its doughy sameness. The search for proof depends, once again, on a plethora of surveys based on interviews and questionnaires, dubious answers to sometimes tendentious ques-

---

* In the context of the Report's treatment of 'satiation effect', 'long-term' is eight weeks.

tions, a vast morass of ambiguity seeking mathematical redemption, passed through the calculating machines, emerging in column after column of percentages and correlation coefficients of admirable precision.

Agglomeration is complemented by its twin, atomisation. Majorities solidify; minorities dissolve. Statistical symbols of individuals are endlessly processed; the individuals themselves never appear *as* individuals, as persons, truly communicated with, as fellow human beings, over time. Time, 'the stream of life', appears as sliced-off moments, tenuously linked by this graph-line. Equally, community, more than ever elusive, defeats the calipers, and vanishes totally from view.

This insistent statistical view of man thus creates a partial vacuum which the rejection of all but numerical proof completes.

At this stage the commercial wing of the Numbers Game takes up the running. 'The evidence of the burgeoning smut industry is everywhere,' reported the *Financial Times* from Washington in March 1971. The Pornography Commission itself estimated the 1969 US turnover in the 'industry' at a minimum of $500m, but thought it might run as high as $2,000m. In Denmark, which had led the way to 'liberalisation', the porno export had become a substantial factor in the balance of trade. In 1963 in Britain, the Secretary of the British Board of Film Censors, Mr John Trevelyan, had expressed the opinion that the wave of 'X' films had 'reached saturation point' and was now 'on the wane'—the famous 'satiation effect' in operation, it might have seemed. Yet by 1970 the British film trade was acknowledging 'Sexploitation' films as 'the one road ... to salvation', and in a special pornography issue of *Today's Cinema*, John Trevelyan in an interview was describing the flood of sex films as 'a licence to print money'.[16]

Not surprisingly, in view of its definitions and methodology, the US Commission concluded finally that it could find 'no reliable evidence' that pornography was harmful.* Or, for that matter, that it was harmless. So, with the aid of that comprehensive sociological escape hatch, 'social function', it accepted the *fait accompli*, recommending that all legal restrictions

---

* With three dissentients out of the eighteen Commissioners.

against pornography be lifted, save for those designed to protect children.*

Although it was, of course, pointed out that Further Research is Needed, it now seemed doubtful whether this would be able to run fast enough to keep up with its subject. For not only are 'licences to print money' rarely turned down, but further research would now encounter that important and versatile mechanism of the Numbers Game, the self-fulfilling prophecy. As it was, the Commissioners were able to claim that research undertaken on their behalf in Denmark showed that a decade of progressive 'liberalisation' (culminating in the 'Everything Porno' shops and Sex Fairs of 1969)† had been accompanied by a decline in sex offences (bearing out the therapeutic thesis).

To the student of the Numbers Game what seems to have happened was much more interesting. Violent sex crimes including rape had neither increased nor decreased significantly. Much of the decrease in other sex offences (making up the statistical bulk) could be accounted for, firstly, by the fact that some offences had ceased to be offences, and, secondly, by the fact that others were no longer reported to the police after the state had made clear its changed view of their gravity by its 'liberalisation' legislation. It is notorious that there is a particularly large 'dark [i.e. unreported and unrecorded] figure' for sex crimes,‡ making statistics highly unreliable, and it is hardly remarkable that there should be a fall in the reporting of, say, 'sexual exhibitionism', when with the consent (which surely must imply some sort of approval) of Authority sexual exhibitionism of every variety was becoming the basis of a booming photographic, film, and 'live show' porno industry.

---

* The question of the most expedient tactics in the social control of pornography is separate from the issue of 'harmful' or 'harmless'. But while the Commissioners did invoke expediency, and the more important issue of freedom, their final attitude by no means wholly rested upon them. Furthermore, it is not immediately obvious how one is going to 'protect the children' in a society pervaded by industrialised pornography with all its ramifications.

† Pornography became generally available about 1965, a situation legalised for the written word in 1967 and for pictures in 1969.

‡ In evidence given before President Johnson's Commission on Law Enforcement, half the complainants who had failed to report sex crimes gave as their reason: 'The police wouldn't want to be bothered.'

Similarly, a decline of almost 80 per cent in the sex offence of 'peeping' may point not so much to porn-fed health as to the fact that it is superfluous to peep through curtains when one can enjoy all the delights of *voyeurism* by gazing into shop-windows or attending skin-flicks. We are all Peeping Toms now.[17]

Although the experts might enter their usual caveats, the statistical 'indicators' that today define reality had spoken. The dynamic 'normality-deviance model' had shifted to a new equilibrium; pornography was the progressive thing now, and in the spring of 1970 the Danish Gallup Poll made the headline that 57 per cent of the Danish people, together with many of the clergy, approved the new 'facts'.

We may be sure that had the splendid but 'invisible' garments of Hans Andersen's Emperor been detailed statistically the child would have cried out in vain that he was, in fact, naked.

8

THE OPIATE OF OBJECTIVITY

The principal categories of social science for classifying human behaviour are 'normal' and 'deviant' ... The truth is the closest statistical approximation to the observed occurrence of events ...

> Professor Daniel Lerner in *Evidence and Inference*, Boston, (Mass.), 1959.

There is nothing more tragic in the present intellectual state of our country than that people have stopped believing in value-judgments—the philosophical belief that there is something to the beautiful, something to the true, something to the good which is not simply a subjective, emotive noise ...

> Quintin Hogg, MP (now Lord Hailsham, Lord Chancellor), Parliamentary Debates, 4 May 1970.

All ignorance, like all knowledge, tends to be opportunistic ...

> Gunnar Myrdal, *Objectivity in Social Research*.

President Nixon greeted the conclusions of the Commission on Obscenity and Pornography with a retort worthy of Dr Samuel Johnson : 'Centuries of civilisation and ten minutes of commonsense tell us otherwise.' It should, but the question remains, why hadn't it ? How can we be so silly, so often ? How can we accept commercialised smut as spiritually liberating or socially hygienic, mathematically processed ignorance as crucial considered opinion, a one per cent change in a dubious statistical abstraction called 'the growth rate' as a supreme indication of national progress or decline ?

The answer that has been suggested is that the mainly statistical evidence leading to these conclusions is accepted as possessing the 'objectivity' of the Scientific Method before which 'subjective' personal opinions, however deeply felt, must bow the head, before which 'obsolete' history is silent, and Johnsonian commonsense is made to seem puerile.

But just how objective is 'objective'? A riddle on which so much of the structure of belief and illusion in our time depends is perhaps better initially met with a parable. One day in 1960 a reader of a London newspaper wrote a letter to the editor reporting on the activities of his village grocer. The man, he complained, was 'making a handsome profit out of the gullibility of some housewives. Each week he buys a new cheese, cuts it in half, and puts both pieces on show, one marked 'Best' at 3/6 a lb., the other at 3s.

'The 3/6 lot sells like magic, but the 3s piece languishes on the counter. So he cuts it in half, prices one half at 3/6 and off he goes again. He does this right to the last piece, and he has hardly ever sold any cheese at 3s yet.'

The baffled reader ('J. G. C. Surrey') concluded: 'There must be a moral somewhere, but I'm blessed if I know what it is.'[1]

There must indeed—in fact 'the gullibility of some housewives' might stand for the gullibility of all of us in face of more complicated, but, basically, not dissimilar, forms of objectivity. The grocer may be considered to have engaged in a rather modern piece of market research, both maximising his yield and meeting his customers' ascertained 'needs'. Since, however, both halves were undeniably the same cheese—objectively, it is hard to see how 'Best' can be other than a highly subjective judgment. Yet the price of Best—3/6—is a market price, and we are not likely to get the practitioners of that most sternly objective of the social sciences, economics, to agree that price set by the market is other than objective.

Nor can we afford to get too involved with the complexities of the parties' motivations, for to retain its authority and sustain its scientific claims, social science must remain, detached and white-coated, on the outside, looking in. It is from the extreme awkwardness of this position, when rigorously insisted upon, that most of the current oddities, the more outrageous 'side-effects', stem. For it means that in order to study our nature we must disavow it; to end with a diagnosis of society we must begin by leaving it.

From this 'natural science' stance, looking down on it through the statistical lens, society inevitably appears as 'the given', ideas appear 'caused' rather than causing; ideology is not 'ended', it merely disappears from view. What *will* be seen will be some sort of functioning system—which statistical analysis and the multiple correlation mill can be trusted to clarify. Sometimes the system appears analogous to that of Nature; sometimes it resembles a machine. Sometimes, as, rather astonishingly, with economics, the metaphors embrace both at the same time. But you can no longer have evil, since this requires someone to experience it, or even the 'unnatural'—because whatever happens has happened—but merely 'equilibrium', adjustment, function and 'dysfunction' and so on. These are words which appear aseptic, scientifically neutral, admirably devoid of the least 'element of subjectivity'.

Thus, given the predominance of social scientists on the Commission on Obscenity and Pornography, the conclusions were predictable before even the first 'blue' film had been shown to the first sample of subjects. Organised pornography is a well-established 'given' of industrial society. The more detachedly and elaborately one studies it (and its 'function') the more one necessarily accepts it as if it were a phenomenon of nature. One social scientist is quoted in the report as saying that 'the half-life of pornography is approximately two to three hours'—as if it were a God-given radio-active substance. In the same way, as in Kinsey, it is very readily assumed that pornography's alleged effect in removing inhibitions is *automatically* good.*

As a smooth, and indeed invisible, method of passing the buck, the 'functional' approach in the hands of social scientists can easily beat any form of words the slickest politician can devise. Thus in contrast to the 'mere impressionism' of 'moralising literati' to use a term recently favoured by a British social scientist,† we have the rigorous objectivity of the CBS Social Research Department, directed by Dr Joseph T. Klapper, who has tended to favour what he calls 'The Phenomenistic (Situational or Functional) Approach'. This brings the thought that 'Mass communication ordinarily does not serve as a necessary

---

* It is, of course, a sign of the times and of current passivity that so many habitually employ the word 'inhibition' exclusively in its neurotic and negative sense, rather than in its equally valid, normal and positive sense, as a part of the mechanism of socialisation and self-control, or even its neurological sense. ('Even a simple activity like walking requires for its performance the co-ordinated inhibition of certain muscle groups in the body'— Hubert Bonner, *Dictionary of the Social Sciences*.)

† James D. Halloran, in *Television and Delinquency*.

and sufficient cause of audience effects, but rather functions among and through a nexus of mediating factors and influences' which are such that 'they typically render mass-communication a contributory agent, but not the sole cause, in a process of reinforcing the existing conditions.'[2] This approach leads naturally enough to 'Use and Gratification Studies' in which emphasis is switched from what television does to the viewer to what the viewer does with television. Thus, the adolescent who appears unduly absorbed in TV violence may well, suggests Dr Klapper (a member of the Pornography Commission) primarily be compensating for lack of success with his 'peer-group' or in school or family life. The real need may be to change the child *before* he comes to television.

It will be seen that as a source of total inertia the possibilities of the 'functional' approach are rich indeed. Thus, among research subjects suggested by the Committee for Research on Television and Children in the US, we find the following: 'Do commercials function as tension-releasers?'

More Objective Than Thou

'Our whole literature,' Gunnar Myrdal has frankly written addressing his fellow social scientists, 'is permeated by value-judgements, despite prefatory statements to the contrary ... They are most often introduced by loading the terminology.' Words, for instance, like equilibrium, balance, stable, normal, adjustment, lag, or function ...[3]

But also, as has been suggested, superlatively by the ever-increasing use of statistical forms and mechanisms which both distance the subject and so perfectly mask the built-in value assumptions, even from the experimenters themselves. The result can be that the conclusions are predetermined from the outset by the choice of 'constants' and 'variables' in the statistical design.

After a generation dedicated to surveying and statistically analysing the factors making for criminal behaviour, the Gluecks in 1968 turned from the scientific prediction of delinquency (through 62 tested tables) to the question of its prevention. In their book *Delinquents and Non-Delinquents* they sum up the conclusions of this latest research. They deprecate those who wish to spend vast sums on housing projects, recreational services and large-scale environmental improvement. While 'the slums should be eliminated (if possible)', they would concentrate on more personal intervention to improve the moral character of family life and straighten out the 'character distortions' which they find to be the critical factor. They want 'to bring all the resources of mental hygiene, social work, religious and ethical teaching' to work on these human deficiencies. Thus, feckless parents would be 'taught the simple elements of dynamism of child-parent relationships'.

The Gluecks have, of course, arrived at this rather narrow conclusion with scrupulous objectivity. But how? The survey which promoted it was designed so that both delinquent and control groups lived in slum areas, both equally suffered poverty and lack of occupational and cultural opportunity. Clearly, it seemed that one must look elsewhere than these circumstances for the factors causing the delinquents to be delinquent, and the Gluecks find them in the moral, rather than the material, poverty of the families. This, all along, had been one of the best 'predictors'.

Within the particular frame of the survey, the conclusion is doubtless 'scientifically' valid; equally, it is the frame selected which determines this emphasis, and excludes others. If the water supply of a city becomes contaminated by bacteria, only a small number of those who drink it may succumb. Many may suffer a slight indisposition, then rally, and become immune. In others again no trouble at all may be *visible*. But although those who succumb may be among the weakest and most ill-nurtured, we should not blame their fate on their weakness. We should find the 'cause' in the contaminated water supply, and look urgently to its purification.

Yet in terms of the Gluecks' survey the water-supply (poverty) would be a constant, even an element in the social norm to which one was required to adjust—or define oneself as delinquent.*

* Commenting on the Gluecks' concentration on personal characteristics and domestic relationship, the British sociologist, Barbara Wootton, writes that the design of the Gluecks' work 'eliminates the influence of external events and of all but a narrow range of socio-cultural factors' (*Social Science and Social Pathology*).

Other social scientists in the field, using different—but equally 'objective'—statistical designs have arrived at very different conclusions. In his Baltimore study, Bernard Lander found a 5 per cent delinquency rate in the best section of the town compared with 55 per cent in the slum area and 35 per cent in a deteriorating neighbourhood. He found he could account for almost all the variation 'in terms of two fundamental underlying factors—an anomic factor and a socio-economic factor.'[4] After examining the data in seventy-four Chicago communities, Belton Fleisher demonstrated a positive correlation between delinquency and unemployment rates,[5] and Professor Ivan Berg, a Columbia economist, told President Johnson's Commission on Law Enforcement that 'the economic circumstances of low income groups multiply many times the probabilities that they will contribute disproportionately to delinquency and crime ... in the inner city the delinquency rates are the highest decade after decade ... regardless of what population is there.'[6]

In such a field as criminology there appear to be almost as many provisional conclusions as there are social scientists. Each claims to be scientifically objective; each aspires to a definitive, proved result. The questions are urgent and the answers are needed now. But the concerned layman, seeking enlightenment from the 'scientists', finds himself in a Byzantine maze with many turns, but no exits.

Whichever the variables selected for correlation, the general effect is still apt to be that so much effort and time goes into 'objectively' classifying and analysing 'what-is', that little energy can be left over to change it—and this might in any case seem vaguely improper as threatening unobjective, and decidedly untidy, involvement. 'What-is', rendered into noncommittal statistics, is thus apt to appear the natural order, and the role of social science must be to accommodate men to it. As modern medicine seeks with tranquillisers and sleeping pills to adjust patients to an overstressful way of life, the Pornography Commission in the United States speculates on the fruitful possibility that 'graded exposure may immunise in somewhat the same fashion that exposure to bacteria and viruses builds resistance ... and the total lack of exposure would render the child who is totally unexposed as helpless as the animal raised in a totally sterile environment.' Conversely, sex offenders (39 per cent) who indicated that pornography 'had something to do with their committing the offence' are accused of 'scapegoating'. Did not one experiment show that finding erotic materials 'disgusting' correlated positively with 'high sex guilt'? The trouble lies deeper; 'in familial and sibling relationships'[7] and in the home, which, inferentially, failed to immunise properly.

Again, the vulgar, non-scientific reader may marvel at the unexceptionable way in which, in effect, the buck has been passed. Having foresworn the unquantifiable insights of history, not to mention moral or value-judgments, social surveyors of media violence may report that children, no doubt 'immunised', no longer betray alarm. But, as Dr Fredric Wertham has pointed out from his New York experience in child guidance clinics, the thing to worry about may not be that the children 'get frightened, but that they do *not* get frightened.'[8] He quotes Dostoevsky: 'The best man in the world can become insensitive from habit.'

And, of course, today's vast captive audiences can be desensitised on an industrialised scale Dostoevsky can hardly have dreamed of. Even so, it seems that there is little that can be done. In a model of the 'Mass Media as a Social System', illustrated by a circulatory diagram of much complexity, a social scientist named Melvin de Fleur reports that 'critics may achieve some temporary disturbance of the system ... and if persistent enough, may even displace some specific form of "low taste content", but that will be about all, because "low-taste content" is the key element in the social system of the media. It keeps the entire complex together ... the financial stability of the system can be maintained.'[9]

Clearly, the role of religion as 'the opiate of the people' has passed on elsewhere. It is no longer 'pie-in-the-sky' that makes possible our resignation, but the vision of an eternal 'Other World' of statistical essence and pure trend lines, beyond current vanities. Thus, as the fathers of the Church, threatened by persecutors, might have turned to prayer, the response of social scientists, under the threat of McCarthyism in America, was to institute a mammoth survey of their own reactions under these stresses. Financed by the Fund for the Republic, this remarkable document runs to 450 pages, and is based on structured interviews with 2,451 social scientists in colleges all over America. Entitled *The Academic Mind, Social Scientists in Time of Crisis*, it contains a wealth of indices, correlation tables and graphs. In order, initially, to select the 'leading social scientists' for the survey a 'rough index of productivity' was prepared, based on the number of publications, ranging down from the top social scientists with impressive outputs to the scorers of 'O',—those teachers who lacked a PhD, had published neither a paper nor a book, and had delivered less than three papers at a meeting.

An 'incident count' was then made—an 'incident' being some sort of accusation, generally political. A classified chart of incidents and outcomes appears in the book. Twenty-one items in the questionnaire were directed to defining a professor's state of mind as a result of such incidents, the data thus processed providing indices of 'Pressure or Worry', of Caution, of Apprehension, of Concern for Civil Liberties, and of 'Faculty Intimidation'. Six of the 21 items indicated 'Worry'; five, 'Caution'. Collated, scored, weighted, they furnished the master Apprehension Index. A teacher, for instance, might be neither worried nor cautious; worried, but not cautious; or worried *and* cautious.

The Apprehension Index was calculated for each type and size of college in the survey. It was discovered that 'the longer social scientists have taught at a college, the less Apprehension they experience.' It was also found, rather oddly, that the more 'cautious' a professor was rated, the more often he 'spoke out'. (It might, of course, have been that owing to his caution, it just seemed that way.)

More than a hundred pages are devoted to examining the validity of the data and methodology. However, when the last correlation is in and has been cross-checked, there remains a characteristic difficulty. How much Apprehension amounts to apprehension? For an objective answer the surveyors feel they need to study the matter over historic periods. Unfortunately, 'we have no such material'. But after analysing many questionnaires, 'it is our impression that two or more out of six items answered affirmatively do indeed indicate apprehension in the colloquial sense of the term. Whether or not this is correct must be left to future inquiries.'[10]

Ironically, this massive, infinitely painstaking, objective

analysis of the American social-scientific mind 'in crisis' under the impact of McCarthyism, was mounted in 1954 just when that fearsome—also statistics-based—'objectivity' was about to collapse under the first determined resistance, the onslaught at the public hearings by the Army's lawyers.

Ivory Towers Inc.

In point of fact, far from enjoying the detached vantage-point their professional stance seemed to require, social scientists were increasingly woven into the fabric of the societies they sought to analyse.

In the book on *Radio Audience Measurement* which he published in 1944, in collaboration with C. E. Hooper of 'Hooper-ratings' fame, the psychologist, Dr Matthew Chappell, wrote with enthusiasm of a revolution 'currently in progress' and encompassing most of the social sciences which would bring about 'the migration of research from the seclusion of the academy to the field of competitive enterprise'.

Business and industry, Dr Chappell went on to point out, had 'developed a fundamental interest in the behaviour of large numbers of people'. It was the 'embryonic stage' of a 'great branch of science—the science of human behaviour in the mass' and only business and industry would be able to supply the funds to finance it on the necessary scale.

Although in what President Eisenhower was to call 'the military-industrial complex' the line between Government and industry was to grow faint, this was to prove a reasonably accurate prophecy. By 1967 Daniel S. Greenberg in *The Politics of Science* was able to write of the scientific community's rapid postwar ascent from 'deep poverty to great affluence ... from academe's cloisters to Washington's high council'. And the Government, industry and the multiplying foundations* nourished not only the natural scientists, but a proliferation of social science specialisms.

---

* By 1952 the total number of foundations in the US—27 in 1915—had reached 4,162, and of these 178 had assets exceeding $10 millions. By the sixties two-thirds of the total funds for research and development emanated from the Federal Government, and three-fifths of scientists and technologists

For the new rewards and status there was, however, a price. By 1970 we find Reinhardt Bendix, Professor of Sociology at Berkeley, writing of 'the declining autonomy of academic scholars, a development which in the United States has coincided with an increase in their general prestige and the increasing availability of extramural funds.'[11]

It was all a part of the industry of 'knowledge production' which according to the President of the University of California, then Clark Kerr, was now 'said to account for 29 per cent of the Gross National Product' and to be 'growing at about twice the rate of the rest of the economy'. 'The university and segments of industry,' Clark Kerr added in his book, *The Uses of the University*, 'are becoming more alike. As the university becomes tied into the world of work, the professor—at least in the natural and some of the social sciences—takes on the characteristics of an entrepreneur ... The two worlds are merging physically and psychologically.'

The industrialisation of the professions was complemented by the professionalisation of industry—the aspirations to Management *Science*. In America in the 1950s it was noted that a 'typical volume of the *Harvard Business Review* had about one-third of its contents devoted to the social sciences'.[12] In conscious emulation of America, the same linkage was being made in Britain also. For the first time 'Business Studies' or 'Management Sciences' now became degree courses in a number of universities and colleges of technology. The Birmingham Graduate Centre for Management Studies, for instance, boasted a research fellow in the 'behavioural aspects of management' and a senior lecturer in social psychology. Linked closely with Midlands big business, the new University of Warwick had chairs in Management Information Systems, in Marketing, in Business Finance, in Industrial Relations and in Business Studies, mainly endowed by large business firms. Again following the American pattern the university looked with favour on its teaching staff holding part-time industrial consultancies 'in connexion with business problems'.[13]

---

in research and development were supported by Government funds—underlining Eisenhower's warning in his farewell broadcast against a state of affairs where 'a Government contract becomes virtually a substitute for intellectual curiosity.'

This was the new orthodoxy, ritualistically confirmed when a long succession of famous British firms and institutions called in the American management consultancy, McKinsey, to run the slide-rule over them to ensure that they conformed to the latest statistical criteria and the 'laws of management science'. Among those thus sprinkled with holy water were the Bank of England, the BBC, Rolls-Royce, the Post Office, the P & O Shipping Line, ICI, British Railways, and the Gas Council: one can only wonder that Buckingham Palace was somehow left out.

The ecclesiastical metaphor seems reasonable enough—for once again the suggestion of a priesthood is strong; it is all here, the arcane language, the doctrinal tests, the handing down of the Word (or in this case the Ratio), penances to be performed, and, finally, the granting of absolution. Here are the guardians certainly not of the 'Divine Order' detected and analysed by that earlier statistician, the Reverend Süssmilch, but of a Scientific or Managerial or 'Efficient' Order equally immanent and no less mandatory.

The Seamless Fabric

In his book on *Motivation in Advertising*, Pierre Martineau, then Research and Marketing Director of the *Chicago Tribune*, shrewdly notes: 'Every society forces its members into moulds of behaviour which they come to accept without question, as if there were no other scheme of things.'

In point of fact, of course, there are, have been, and just possibly may yet be, many other 'schemes of things'. One of the most striking things about man is his malleability. He can delight in the thrush as a songster—or see it as something to shoot and eat for tea with equal conviction of the rightness and normality of this behaviour, which derives from the conditioning culture. American-style capitalism or Soviet-style communism do not—although to listen to some one might think so—exhaust the possibilities of life.

It was of course, of the former, or of what we now call 'the consumer democracy', that Mr Martineau was writing from his unrivalled vantage-point in Chicago. And certainly no society better bears out his observation. Paradoxically, at a moment when—through technology with its immense range and the attendant social sciences—our ability to change is greater than ever before in history, the sense of inevitability, of *as if there were no other scheme of things*, is stronger than ever. The ordinary man is flatteringly confronted with a greater *apparatus* of choice than ever in the past. The opinion pollster, the market research lady, the earnest young social scientist are for ever hovering about our doorsteps, soliciting our views. Yet even as we answer we have the dispiriting feeling that somehow the choice we are being invited to make has already been made. Even the young, triumphantly 'doing their thing', turn out to be conforming rather minutely to somebody else's thing, down to the passwords uttered and the carefully-carelessly cut-off faded jeans. According to Mr Alvin Toffler, 'the successful "sale" of the hippie life-style model to young people all over the techno-societies is one of the classic merchandising stories of our time.'[14]

It is, indeed, precisely the multiplication and elaboration of the new mechanisms of choice which deny us choice, because their complexity and appearance of 'scientific' objectivity conceal what is really happening, and fob us off with the illusion. The real sources of the drive built into the mindless synchromesh are concealed from view. The one-time 'profiteer' of simpler days, maximising yield on assets employed at whatever cost in social and human disruption and waste, becomes the unassailable practitioner of Management *Science*.

And now this canopy of 'objectivity' is extended over a larger area of our society. At the turn of the century the Fabians in England set out to infiltrate society and particularly its governing élite with socialist ideas. In fact, an infiltration of a very different character has taken place, unavowed, not of an overt ideology, but of something much more treacherous, of an 'objective' acceptance of the 'given'—'as if there were no other scheme of things'.

It is a process which has been much forwarded by the close interweaving, through the mechanisms already noted, of some branches of social science, of commerce and technology and mass communications to form the seamless fabric of contemporary society. Increasingly, as already suggested, this continuity is to be found in persons. Thus the case of Dr Frank

Stanton, in America, the social psychologist who became President of CBS, a trustee for the Center for the Advanced Study of the Behavioural Sciences, one-time Chairman of the Rand Corporation and so on—easily spanning the worlds of research, mass communications and big business—is mirrored in minor key—the time-lag?—in England in the role of such a figure as Dr Mark Abrams, social scientist, leading market research man, opinion pollster, former President of the World Association for Opinion Research, and director of a leading British advertising agency.

When the old Labour daily newspaper, the *Daily Herald*, was to rise again as the bright, modern *Sun*, the publishers commissioned Dr Abrams to prepare a detailed profile of the newspaper reader of the next decade and his aspirations. (Once the product of creative editors, modern newspapers and magazines were now increasingly designed in this fashion—as indeed were political parties, in which process also Dr Abrams played an important role.)* The publishers of the *Daily Herald* were thus, after objective research, advised to bear in mind that the activities and interests of women would be the main shaping force in the family's ties with the outside world. Political outlook would be 'secular'. 'Ideological politics would be displaced by consumer politics and political support given to those who had offered the greatest competence in social engineering.'[15]

It will be noted that, with the usual social scientist displacement effect, these developments are presented as, in effect, self-creating, the product of historic forces, almost predestined. Who would guess, looking down upon this seamless web, that some of the most vigorous and persistent 'social engineering' is in fact being continually engaged in by the advertising, market research and marketing industries of which Dr Abrams is himself so distinguished a representative? Or that this social engineering is directed towards encompassing, to its own not inconsiderable profit, the secular developments the statistics objectively report?

And it isn't even necessary, in fact, to change hats between roles because market research itself in its various ramifications

---

* In surveys cutting a statistical template, so to speak, for a 'modernised' Labour Party: see chapter 10.

—analysis of changing class structure, motivational research and so on—is equally enfolded in the white coat of social science, and its disinterestedness thus placed beyond question. Market research, rules an Anglo-American authority, 'is the way in which the seller reaches an understanding of his customers and optimises the operational behaviour he engages in to meet customer needs.'[16] With this emphasis on serving the people's needs, market researchers may indeed claim to be close to the throbbing heart of the 'consumer democracy'—to be in the vanguard of that Permanent Revolution which the editors of *Fortune* in 1951 declared to be truly naturalised in the United States.

However, closer study of the texts reveals a redistribution of emphasis. Market research is the tool of marketing, which is defined by Ralph Glasser, a British marketing consultant, as 'the skill of selecting and fulfilling consumer desires so as to maximise the profitability per unit of capital invested in the enterprise.'[17] Particular attention should be given to the verbs 'select' and 'maximise'. Of the notion that 'marketing research ... is an indication of the large corporation's concern with the needs and wishes of the consumer,' another highly experienced British marketing research director and 'product-planner', Graham Bannock,* declares roundly, 'It is nothing of the sort ... The amount of resources, intellectual and financial, devoted to finding out what the customer wants or needs are very small indeed ...' Much of the money and effort goes on monitoring the market, establishing brand shares and so on, and 'almost all the rest is concerned with testing marketing methods, that is, trying out things on the consumer that the corporation wants to do ...'[18]

And Bannock points to a fascinating change in the British Institute of Marketing's definition, made in 1966. The earlier definition of marketing included 'assessing customer needs and initiating research and development to meet them'. But in the revised version reference to 'customer needs' has vanished. Marketing is now directed towards 'assessing and converting

---

* Among other positions, Graham Bannock has been market research man at Richard Thomas and Baldwin, the steel firm; Head of Economic and Market Research at the Rover Motor company; Manager of the Market Research Development at Ford, Great Britain; and Manager of Advanced Programmes for Product Development at Ford of Europe Inc.

consumer purchasing power into effective demand for a specific product or services and ... moving the product and services to the final customer or user so as to achieve the profit target or other objective set by the company.'

'Converting consumer purchasing power' ... 'to achieve the profit target'—we are now getting nearer to the built-in over-mastering drive in the system, the compulsion to ever-greater statistical yield, to the annual profit growth as the supreme and indispensable symbol of virtue and 'efficiency', the mindless, tyrannical numbers that all minds must be bent to serve. In fact, in this service, as Bannock, Glasser and many others have pointed out, far from being directed towards enlarging true consumer choice by multiplying genuine options, marketing research as dominated by today's large corporations is very often directed towards making it possible to narrow the pro-duct range—thus boosting profitability—by evolving effective ways to substitute the illusion of variety for its reality.

Such concentration on packaging gimmicks, superficial style changes, differing brand 'images' and so on not only fails to widen true choice but all too often has the effect of losing even simple quality, so that the inhabitant of some 'primitive' village may fare better than the citizen of the 'techno-society'. In Britain, for instance, it has never been so easy to buy a vast range of picturesquely named brands of bread, 'Mother's Own', 'Crusty', etc. etc., but never more difficult to find, at a normal price, any bread not denatured and without taste, while in America, the archetypal consumer democracy, the *Washington Evening Star* was recently moved to denounce, in an editorial, the 'miserable' consistency of the American white loaf, com-plaining, 'Hardly anywhere on earth could non-Americans be forced at gunpoint to eat what we call bread in this country, with the possible exception of Biafra and Bangla Desh.'

Now the same sad fate overtakes ice-cream, even in the land of its birth, for Mr Motta with all his immense output and nationwide refrigeration and glossy, gaudy showcards, does not deal in the same delectable commodity which the old hokey-pokey man brought, no doubt unhygienically, to our northern climes.

It would be easy, but it is unfortunately quite unnecessary, to multiply instances. Nor is it necessary to embark on a philosophical discussion of the 'reality' of the 'market-researched' choices in fact offered—upon whether, for instance, the soap which 'Creams Your Skin While You Wash' (which according to the research of David Ogilvy pulled 63 per cent more orders than the next best promise)[19] is really better than the same soap which just lathers; upon whether the identical beer becomes 'better' when its price is raised and it is advertised as 'premium';[20] or even whether meat pies, with virtually no meat, are What the Public Want (because they have so long been getting)—an issue which occupied the Food Standards Committee of the British Ministry of Agriculture, off and on, a good seven years.[21]

It is unnecessary because we are not yet all Walter Mittys, content to live on a diet of fantasy. But for how long? Much-touted campaigns for factual truth in advertising cannot save us. Because under the guidance of sophisticated market research this is an area where verifiable specifications figure minimally, and statistics serve merely to provide directional guidance for the balloons of fantasy. 'Attention-getting, competitiveness, sex, hunger for praise, power drives; these are just typically important motives which can scarcely be approached except by indirection,' points out the practical market man, Pierre Martineau. 'That is why the most effective advertising, like successful public relations, is the art of saying one thing while the important meaning is often sailing out in another direction to appeal to the real motive areas.'[20]

It is not without significance that market research, the intelligence branch of this curious, complex, guerrilla war for the possession of the consumers' pence and fancy is today a fast-growing industry in its own right. 'Market research is expanding so rapidly,' reported the London *Times* in 1962, 'that last year the net of interviewers captured one adult in every ten of the population.' Britain has still some way to go to catch up with the United States where marketing research is estimated to have increased twentyfold in the last twenty-five years and now grosses over £200m annually.[22] Nevertheless, when the television-and-consumer 'audit' firm, AGB Research Ltd, was floated on the London Stock Market in January 1970 its share price doubled in four weeks.

Thus the pivotal position occupied by such statistical monitors was recognised. On one side, market research stretches out owards the Press, and is organically linked to it by the need to shape papers to fight successfully for limited advertising revenue. On the other side, it joins hands with the opinion polls, generally run by market research firms, for whom they normally provide small revenue, but much publicity. Again, the fabric is seamless: the 'pulse of democracy' beats so close to the arteries of commerce that one might reasonably anticipate some sympathy of rhythms. The headlines prompt the questions—and answers and the polls make more headlines. Technically, the pollsters need clear-cut questions yielding clear-cut, addable answers. So do the headlines.

When to the difficulty of including the unquantifiable is added the gravitational pull of what management science would call 'profit centres', the polls' obsessional concern with 'the standard of living' ceases to surprise. Nor is it remarkable that such a publicised, quantified—moving Up? moving Down?—'standard of living' should be found in Britain to be 'the strongest long-term influence on voting decisions', or in America to form a social and material imperative to which, as Jacques Barzun has observed, it becomes necessary to conform 'under pain of being cut off from the experience of the race.'[23]

New Sins for Old

While all social scientists are, of course, objective, some are more objective than others. But none are more objective than those who are most continuously and legitimately concerned with numbers, dedicated as they are to that supreme objectivity, price. In the exclusion of value-judgments none are more dedicated; of the many equilibria celebrated at the altars of the great cathedral of statistical science, none commands the authority or inspires the awe of the economists' Equilibrium. And although few economists would today go along with Malthus, the first professor of political economy, in hinting that economic laws are of divine origin, nevertheless no influence is so brilliantly effective in inducing members of society to accept the 'moulds of behaviour' into which they have been forced *'as if there were no other scheme of things'*.

It is a remarkable achievement in so short a time. For eighty or ninety years ago it still seemed that certain options were open. Studying the 'new economics' in the works of her friend, Professor Alfred Marshall, that sharp-eyed woman, Beatrice Potter (later Mrs Sidney Webb) noted that although this 'self-contained science' purported to state universal laws, it was in fact merely a description, or abstraction, of one particular variety of 'profit-making capitalism'.

And she dared to go on to say that, in her work as a social investigator, she had observed that men do not in fact pursue their 'pecuniary self-interest' in the consistent manner they were assumed to do in 'the mechanistic doctrine of orthodox economics' as favoured by her friend, Professor Marshall. Pronouncing the 'delimitation of content' 'misleading', she asked who would 're-make the science of economics'[24] in a way that would take into account the true width of human motivations and other possible modes of organisation, such as socialism.

Alas, it was already too late. The anaesthesia of Objectivity was already taking effect. Soon, when pronounced by ordained economists, accusations of 'distortion' (implicitly from some 'natural' economic order) had become as grave an indictment as 'deviation' in the Communist world, and as shameful as mortal sin in Catholic circles.

It was not, as Beatrice Potter had hoped, the socialists who were to re-write economics, but the economists—particularly the 'socialist' economists—who were to re-write—and write off—socialism.

9

HANDS—INVISIBLE, VISIBLE AND DEAD

... to think of economic policy in terms of engineering has potentially dangerous results ... the search for the philosopher's stone is a recurrent motif in the history of economic thinking and it leads nearly always to excesses which give anti-rationalism a chance to extend its hold over men's minds.
> Sir Eric Roll, *A History of Economic Thought.*

Through the development of mathematical and statistical theory, through econometrics and the techniques or empirical testing, we have come to realise how little earlier economists really understood. We need no longer rely on commonsense; we have unerring scientific criteria.
> Professor Ely Devons, BBC talk, March 1963.

Economics is a very dangerous science.
> J. M. Keynes, essay on Malthus, in *Essays in Biography.*

Just as there is no business like showbusiness there is no 'indicator' like an economic indicator; in sheer quantity, authority and power of intimidation they are without rival. Combining the utmost mathematical precision with the most generous embrace, economic indicators today stamp our lives and times with their values like price stickers in a supermarket.

This curious and all but total domination was vividly illustrated on both sides of the Atlantic in the summer of 1971.

Britain confronted what her Prime Minister, Edward Heath, called 'a historic challenge ... the chance of new greatness'— the decision whether or not to 'join Europe', otherwise the European Economic Community. 'We've faced many great challenges in the past,' Mr Heath told his television audience, 'but no one can say we were found wanting.'

Seeming to portend the end of Britain's long-cherished island status, on which so much of her national character had been founded, the moment was certainly not without its resonance.

But the promised 'Great Debate' was, in the event, conducted in almost wholly numerical terms, a weary and protracted wrangle over the probable course of the economic indicators, average wage rates, balance of payments, investment ratios, market capacities, food price indices ... It was more like an accountants' convention than a nation engaged in reaching what was—or should have been—an essentially political decision on the sort of life and society it sought to achieve. Above all, rising above the clatter of the rival calculating machines, was the swelling organ theme of the Growth Rate, the fugue of decimal points gained and decimal points lost, the ranking order of 'standards of living', conceived as Gross National Product per head, leading to that haunting English affliction known as 'lagging behind'.

Thus the British demonstrated how completely they had entered that economists' Other and Better World of statistical fantasy. For apart from the spurious precision of the Gross National Product figures, and the extreme difficulty of international comparison in money terms across exchanges which are in different ways and degrees manipulated, the cross-cultural, cross-climatic and cross-national—not to mention cross-personal—ordering of 'standards of living' partakes of the mystical. How *does* one compare, with such authority and precision, pasta and Chianti under the Italian sun with, say, fish and chips on a rainy day in Manchester, eaten while waiting in a queue for the Halle Orchestra?

Never mind. The index numbers are a feast in themselves, and there appeared every prospect that this 'moment of decision in our history'—to quote Mr Heath—would be determined by calculations hardly less abstruse than those of the medieval Schoolmen disputing the number of angels that might be accommodated on the point of a pin.

American experience suggests that, far from bringing the prospect of relief, increasing affluence merely brings a multiplication and even greater tyranny of the 'indicators'. That same summer, in the wake of the recession of 1970, Americans watched the wavering lines on the economic charts with the anxiety of a convalescent heart patient watching his own electrocardiogram in the making. They hung on the quarterly quiverings of the GNP, hopes soaring with the hesitant rise of the monthly industrial production index, only to be dashed by the trade deficit figures, the unemployment index, the GNP price index, the consumer price index, the index of wholesale prices of industrial commodities the business confidence index and, above all, the Dow Jones, for ever promising, and for ever relapsing around its critical levels.

When to all this was added the—contradictory—public interpretation of the indicators by a great variety of economic experts, the predictable result was the characteristic numerical neurosis of 'advanced post-industrial society'. On 28 July, for instance, a drop of 0.5 per cent in that mystical entity known as 'the composite index of leading indicators', following seven monthly increases, was enough to precipitate a substantial fall on Wall Street. Confidence was fragile. In early November a financial correspondent was reporting: 'There are still the hopefuls, hunched over their charts, plotting every upward flicker of the index that might bring encouragement, but overall it is a sorry atmosphere ...'

Sorry indeed!

The Economic 'Mystery'

There was a time, not so very long ago, when economics teachers used to explain that Economic Man was but a convenient theoretical abstraction. But these days we are all economic men. Other social sciences, which one might have hoped would offer a supplementary view, have shown a tendency to ape economics, hoping to gain respectability in this statistical world by adopting its models and terminology; behavioural psychologists work on the human mind, for instance, in 'input-output' terms.

'The economic factor looms so large,' said the ecologist, F. Fraser Darling, recently, 'that people in power use it as if it were one of a Trinity with God and Satan.'[1] The theological note is apt—for economics, in fact, started out as a branch of moral philosophy,* and has been pushing out from within ever since,

---

* Adam Smith occupied the Chair of Moral Philosophy at Glasgow, and taught the subject in four parts: Natural Theology, Ethics, Jurisprudence, and Political Economy. The Revd Thomas Malthus, who was Britain's first (1805) Professor of Political Economy, also approached the subject via the same 'moral' route.

until now it has taken over the whole area, including the power of prophecy and the issuance of 'signs'.

Here, in fact, is the prime source of our contemporary sense of 'inevitability' and powerlessness—in the economists' persistent view of society as a largely self-regulating machine, a concept of 'automation' and 'feed-back' which long predates the computer, but is now complemented by it. Adam Smith with his concept of Equilibrium and of the regulatory Invisible Hand, by which self-interest resulted in the common good; or Malthus with his own grim, self-regulating model, are obvious examples. In later years, business cycles were—in the words of Professor Wassily Leontief—'recognised as quasi-mechanical and automatic phenomena formed and operated to a large extent beyond the calculations, outside the control of, and mostly against the wills of, the millions of individuals whom they affect.'[2]

Ironically, it has been precisely the attempt to break out of this determinism, to intervene in the inexorable processes of the trade cycle—and substitute for the invisible hand of Providence the visible hand of the economist—which seems to have snapped and bolted the door on our squirrel-cage, gilded though it may now be. (This is possibly because the most accomplished econometrician and his model do not yet command the confidence once reposed in the Almighty.)

The story of the impact of Keynes's *General Theory* both in America and Britain has often been told. Coupled with the formulation in America of statistical techniques that made National—or Social—Accounting possible, it moved economics away from the niggling niceties of the price-mechanism to aggregates and the intoxicating sweep of 'macro-economics', and it did so just in time for the manipulation and control of entire economies to be tried out, and apparently proved in face of the all-out demands of total war.

In the postwar years the effect of this 'Keynesian revolution' and of 'full employment' policies then proclaimed was to install economists and their attendant statisticians in positions of influence in the administrative apparatus of the Western nations. With the Commissariat General of Planning, set up in January 1946, France wholeheartedly embraced the new mystique; in Britain under Sir Stafford Cripps the old moralistic socialism of the British Labour Party gave way to managerialism and productivity indices, while even in the United States of America, the last stronghold of the 'Invisible Hand', the visible hand was more and more in evidence.

Yet the types of control were now strictly limited; the vision that dominated the scene was, in effect, that of a finely adjusted, fast-running productive machine. The politicians who had once steered 'the Ship of State' with considerable panache, now found themselves attendant upon a semi-automated container-vessel, steered by computerised controls linked to 'economic indicators', and merely requiring from the Economic Pilot the occasional adjustment, the 'touch on the tiller' to keep her 'steady as she goes', to quote a metaphor favoured by the British Chancellor of the Exchequer, Mr James Callaghan, shortly before Britain's 'crash' devaluation of 1967. Across the Atlantic the yet more advanced vessel taken over by President Nixon in 1968 scarcely required even that 'touch on the tiller'; with Pilot Milton Friedman at the new captain's elbow the vessel's intricate mechanisms were to be controlled through the single lever of 'money supply'.

If the vessels so navigated appeared to spend a remarkably large part of their time heading for the rocks, this was felt to be no reflection on the system of navigation. The charts were scientific; it was the rocks that were wrongly placed.

For the charts were the heart of the Mystery, culmination of that union between statistical science and economics which had grown more fecund—or at least prolific—with every decade since the eighties, when the economist Stanley Jevons had declared that 'our science must be mathematical because it deals in quantities.' And, by the 1960s, with the need for fine adjustment of the economic 'machine' which was society, the econometricians had come into their own. When the first two Nobel prizes in economics were instituted in 1969 both were awarded to pioneering econometricians, Dr Ragnar Frisch of Norway and Dr Jan Tinbergen of Holland, builders of 'dynamic models for the analysis of economic processes'.

The Mystery grew more potent. To quote an historian of economics, Sir Eric Roll, the mathematical techniques of some econometricians now became 'so highly refined' as to 'remain inaccessible even to many fellow economists'.[3] Nevertheless, textbooks on econometrics poured from the presses. The student of economics could now escape at speed from the grimy, untidy world of actual industrialists and salesmen to that Other World of algebraic equations, Greek letters, and econometric models, which grew steadily in refinement and complexity.

In 1936 Tinbergen had published the first econometric macro-model ever constructed, designed to trace the effects of alternative measures of economic policy on employment and the balance of payments. It contained 24 equations. Twenty-nine years later, the Brookings short-term forecasting model of the United States economy contained some 219 equations and incorporated seven production sectors.[4]

This ever-increasing sophistication and complexity represented an heroic attempt to incorporate within the model's array of constants and variables all the contingencies bearing upon the economic situation; to validate the econometricians' claim that their techniques have at long last unified economic theory and practice and made possible, as Dr Ragnar Frisch wrote in the first number of *Econometrica* in 1933, that 'constructive and rigorous thinking such as has come to dominate the natural sciences.'

It is the remarkable success they have achieved in getting politicians, public and press to accept, and even stand in awe of, such claims that has made the economists pre-eminent—so far—among social scientists. It may be attributed not a little to the magic of number. A former Economic Adviser to the British Cabinet, Michael Stewart, has related how in an attempt by economists to show that a higher level of unemployment was needed to curb inflation, 'a series of complicated econometric models was shoved under the noses of ministers and civil servants who, perhaps because they were unable to understand them, were visibly impressed.'[5] 'Phillips Curves', plotted from the data of the past, 'scientifically' indicated the minimum of unemployment needed to keep prices and wages steady. The future was to wreck this, and many other, authoritative predictions. Yet in 1970 *Time* magazine still found it a reader-attraction to acquire its own high-powered 'Board of Economists' to add their Delphic utterances to those of America's already numerous colleges of economic Augurs.

Hooked on Number

This is all the more remarkable in that the econometricians' record in prediction has been by no means as impressive as the volume and complexity of their labours. One check showed 30 per cent of the forecasts made with the aid of the Dutch models wrong and 45 per cent of the Scandinavian forecasts wrong. Another evaluation of six American models, undertaken by H. O. Stekler, of the Berkeley campus of the University of California, showed all unreliable in at least some period or department.[6] A forecast which is right *sometimes*, but one does not know *when*, is a decidedly dubious asset. Yet 'accuracy can in the future be improved only very slowly and to a constantly diminishing degree,' gloomily concludes Professor C. A. van den Beld, Head of the Netherlands Central Planning Office. 'The econometrician will have to reconcile himself to falling short of the precision of science.'[7]

The truly extraordinary thing is that anyone should ever have supposed otherwise. To complete a large-scale model of an economy may take ten years; in that time at least some vital relations built into it may have changed. If the econometrician therefore resorts to a greatly simplified model, *ceteris paribus*— other things being equal—that magnificent economists' escape clause—becomes more than ever vital. Unfortunately, in actuality other things are so rarely equal. 'However carefully formulated,' writes Professor C. E. V. Leser, builder of a forecasting model for the Irish Republic, 'the equations of any economic model are likely to be incomplete and we can never be sure that all relevant variables are included.'[10]

Another authority, Professor Streissler, goes further. He points out that even before they are inserted into the model, some of the indicators have themselves undergone complex and protracted processes of production from the raw statistical data (itself officially adorned with many pendants of reservation). Sometimes the processes involve as many as six disparate stages —each with its own possibilities of error. 'It is,' he writes, 'not at all uncommon that the econometrician at the final stage more or less completely misunderstands what the statistics he works with really mean.'[7]

We have been conditioned to take with the utmost serious-
ness such announcements as that (in March 1970) 'President
Nixon's Administration expects only a 1·3 per cent Growth
Rate this year, along with 4·4 per cent inflation,' while in
Britain much impassioned argument has raged over the exact
Growth Rate to be encompassed: first there was the great
national cynosure of a 4 per cent Growth Rate; then (1964 Plan)
a 3·8 per cent Growth Rate; and then, in 1970, in the 'Task
Ahead' prospectus, a 3·25 per cent Growth Rate.

Each decimal point change was solemnly debated, filling
miles of newsprint, as if the figure represented the hardest of
realities, attainable by adjusting the appropriate throttle on
the 'economic machine'. In fact, as suggested in Chapter 1, it
would be hard to imagine a quantity more amorphous, labyrinthine, miscellaneous, haunted by ambiguities than the Gross
National Product from which the Growth Rate, that neat little
figure in which we are taught to repose all our hopes, derives.
Professor Morgenstern, Director of the Econometric Research
Program at Princeton University, has demonstrated, hair-
raisingly, that assuming for the GNP a not unreasonable range
of error of 5 per cent, plus or minus, this can play havoc with
the calculation of the Growth Rate over two periods, resulting
in rates of 1·8 to 12·5 according to the particular combination of
error eventuating. Not surprisingly, he suggests 'very serious
doubts about the reliability and usefulness (in the current sense)
of growth rates.'[8]

But far from being given pause by the pitfalls of estimating
and, even more, of forecasting changes in the GNP of a single
country,* economists cheerfully go on to project *comparative*
growth rates for all the 24 greatly varied countries of the
Organisation for Economic Co-operation and Development—
and to draw suitable morals. The British National Institute of
Economic and Social Research, for instance, forecast a growth
rate of 2·9 per cent for the whole OEEC area, including the USA
and Canada, a breathtaking exercise in econometric precision
the audacity of which one can hardly refrain from admiring.

* In the article on Prediction and Forecasting in the International
Encyclopedia of the Social Sciences, Victor Zarnowitz of the National
Bureau of Economic Research, New York, comments that 'despite the
widespread use [of such economic forecasts] ... surprisingly little has
been done to test' their results. One wonders why.

Or is it simply that these days 'Dutch courage' has been superseded by 'computer courage'?

In the second edition of his revealing work, *On the Accuracy of Economic Observations*, published in 1963, Professor Morgenstern expresses his disquiet at the way in which demonstration of the shortcomings of many economic indicators leaves even professionals indifferent. 'The textbooks on national income and macro-economics,' he writes, 'show little if any evidence of awareness of these difficulties and limitations. The trade journals likewise go on accepting statistics at face value ... in Great Britain, as in the United States and elsewhere, national income statistics are still being ... interpreted as if their accuracy compared favourably with the measurement of the speed of light.'

In Britain Professor Ely Devons, a specialist in economic statistics, has been as devastatingly frank as Professor Morgenstern in America. He writes of the professional custom of making an obeisance in the direction of 'the error in the statistics' and then 'proceeding blandly as if there were no error in the figures at all. This attitude to error and unreliability,' he continues, 'is implicit in most econometric analysis of the determinants of general economic activity and in forecasts of inflationary and deflationary gaps ... There has been much theoretical examination in recent years of the meaning and significance of what we measure in index numbers, and most of this analysis leads to the conclusion that, except in very special and unusual circumstances, it is not possible to give any significant meaning to index numbers of real national income, production or price.'[9]

Why are we, the public, given so few hints of such monumental uncertainties? There have been two basic reasons for this massive imposture, this international reign of bogus precision, the great economic con game of which we—and it would seem economists too—are the victims. The first is that the power and prestige of economists in our society depend upon their claim to exactitude; without it the global temple of mystery in which they officiate would be reduced to a quite small office; their subject would shrink disconcertingly. The second reason is that, like many another profession, they are persuaded by their own mystique; there is no evidence that statisticians are finally any more proof against the magic of numbers than the rest of us. On the contrary, it is said that love is blind.

Again Professor Devons is refreshingly frank: 'We have all acquired the very convenient habit of keeping our ideas in two strictly divided compartments. We all know the theoretical objections and limitations, but when it comes to using index numbers we manage, Freudian-like, to forget all these. Preferring to believe, if challenged, that for the particular use we have in mind, the theoretical limitations and qualifications can be ignored.'[9]

The consequence of all this, however, may be that the—unstudied—psychological 'side effects' of our proliferation of economic indicators may well be more extensive than their intended effects. Indeed, the contemporary alternation between precise, much-headlined, 'scientific' figures and their subsequent abrupt withdrawal might have been devised as a technique for producing disorientation and undermining self-confidence in a prisoner-of-war interrogation centre. Thus, forecasting Britain's balance of payments for the second half of 1968, the National Institute of Economic and Social Research projected plus 90 in February, minus 40 in May, minus 125 in August, minus 192 in November.[10] When the Old Testament prophets issued their 'signs' they made a brisker job of it.

Furthermore, the spurious precision of the indices, suggesting a knowledge which is not in fact possessed, may distract attention from real causes, encouraging the treatment of the symptoms, rather than the 'whole man'. A productivity index is so much tidier and safer than an examination of the human complexities of man-and-work situations. It isn't difficult to see why statistical artefacts are the favoured rhetoric of this technological age.

So statistical rationalism terminates in the policy of obscurantism. In contemporary Western society the economists and econometricians resemble the high priests of ancient Egypt, possessing the papyruses on which were inscribed the secrets of the rise and fall of the Nile. Just as the fellahin awaited the life-giving Nile Food, so it is our lot to wait for 'the economy' to deflate, reflate, inflate, take-off, slide ... Although in our case the economist-priests themselves do not seem immune from a certain bemusement before the semi-mystical automated processes they are believed to control. 'There may be a certain orbit speed that you have to get in order to make recovery self-sustaining,' mused Arthur Okun, former chairman of the Council of Economic Advisers, recently, providing guidance for the readers of *Time* magazine in June 1971.

In such circumstances, the public—whose co-operation and initiatives are surely not expendable—is naturally reduced to the role of *voyeur* or statistic-gazer—or to that of the established hypochondriac who has abandoned himself, body and soul, into the hands of the doctors.

The Only Direction is Up

This by no means exhausts the contribution of the economic doctors to our iatrogenic neuroses. In addition, by the great persistence with which they have concentrated our attention, every week of every year, on the meter of the production of goods, measured solely in terms of their money values, and by their insinuation that there is something badly amiss with us unless this pointer is continually rising, they have reduced us to the condition of a dog chasing its own tail. Most dogs, however, having reached adult years, give up this exercise. We can't.

It seems odd now to recall that the purpose for which the British and American Governments embarked on the control of their economies was the unequivocally humane one of ensuring full employment, as declared in the British White Paper of 1944 and the American Employment Act two years later. Because by the sixties, the rate of unemployment had become just another 'indicator', a politically sensitive one, certainly, yet for all that merely another variable in the equation leading up to that king of indicators, that supreme, if curiously abstract, goal of all modern man's effort, the Growth Rate.

Civilisation was now commonly presented in the public prints as a function of 'Growth', the virtue of nations determined by their position in this new interpretation of the phrase, 'the human race'. Thus, through this period Britain was required to hang her head, not for any defects in her society—and Heaven knows there were plenty—but for the shameful offence of achieving a mere 2·6 per cent growth rate; whereas Japan, with a 10 per cent growth rate less than a generation after Pearl Harbour—which certainly displayed a precocious grasp of cost-effectiveness—was rapturously hailed as a shining example to all.

'Do we accept a world in which the average Japanese will within a few years be better-off from his own efforts than the average Englishman?' Mr Reginald Maudling, the Conservative Administration's Chancellor of the Exchequer, demanded of his fellow-countrymen in the *Times* in 1968.

In common with many people at that time Mr Maudling did not find it necessary to pause to ask his fellow-countrymen whether they would also accept cities with rudimentary sewage systems (only 40 per cent of the houses in metropolitan Tokyo and less than 10 per cent in the suburbs are connected to public sewage pipes), where gross atmospheric pollution is killing the trees and causing massive respiratory illness, where one has to incinerate one's rubbish in the backyard, and the rivers are filled with domestic and industrial wastes, where dysentery and sleeping sickness are prevalent and hospitals often primitive and few, where over the country as a whole the road system is exiguous and much of it unpaved, and the toll of fatal accidents is almost as high as in the United States with twice Japan's population?[11]

He might also have asked whether the British worker would really consider himself 'better-off' with the submissive spirit of his Japanese counterpart, or under the paternalism of the big Japanese corporation. Or indeed better-off *as* a Japanese! But as Professor J. K. Galbraith has observed: 'St Peter is supposed to ask applicants *only* what they have done to increase the GNP.'[12]

Thanks to a few economists like Galbraith, this particular neurosis is now being increasingly recognised. Indeed, next to 'puritanism', 'Growth' may now be the favourite object of voguish disapproval. In England Prince Philip has publicly excoriated the policy of pursuing 'Growth at any cost', and in his January 1970 State of the Union message, President Nixon called for a 'national growth policy' designed to foster not just growth, but *'balanced* growth'. 'The answer,' he said, in a notable echo of Galbraith, 'is not to abandon growth, but to re-direct it' towards 'a new quest, a quest not for a greater quantity of what we have, but for a new quality of life in America.'[13]

It is a fine aspiration, but as some American statisticians have reminded us, words—these days—are pretty feeble things compared with numbers. And when economists and our new 'numerate' élite of scientific managers and accountants use them, the authority and 'neutrality' of numbers reach new heights. It is not only that, at national level, that extraordinary aggregation, the GNP, is so magnificently undiscriminating, heaping in everything that has the single 'quality' of a money value, missiles and health services, packaging and garbage removal equipment, cigarettes and radio-therapy apparatus for the treatment of cancer and advertising to promote the sale of more cigarettes; it is not only that it and the growth rate tell us nothing about the way the economic product is distributed, either between man and man or between private and public uses; it is also that it is the logical crown of an all-pervasive economic Numbers Game in which the inability to discriminate other than in money or quantitative terms is basic. It is not only that the nation is bound to the wheel of 'growth', but every company with the same inherent lack of discrimination must increase its production and profits from year to year, as a 'successful' newspaper must increase its circulation; and that at the heart of this semi-automated machine the 'institutions' controlling the distribution of investment capital apply, willy-nilly, the same 'blind' statistical 'growth' criteria (although no doubt with the odd act of irrational 'moral' discrimination, such as steering clear of the white slave traffic).* What else indeed could they do when by 1963 the British insurance companies and pension funds had to invest somehow £800m a year while in America insurance industry investment in the stock market had grown from $270m in 1951 to $650m ten years later? Meanwhile in both countries mutual funds and unit trusts proliferated like rabbits on the confident promise, vividly illustrated by the rising graph line in their advertisements, to multiply year by year John Citizen's 'nest egg', so that the 'little

* However, such anomalous behaviour may not continue. According to *The Wall Street Journal*, a West German company, Kohls Liegenehaften with an equity capital of $2,580,000 (to be quadrupled by end 1973) offers a guaranteed minimum return of 12 per cent on its 'units' which are being actively sold to investors through investment consultants and its own salesforce, including some former IOS Mutual Fund salesmen. The company's business: a chain of apartments for rent to prostitutes.

man' might feel that he too in his modest way might echo the feats of statistical levitation of the property developers and takeover and conglomerate kings whose brilliant wealth-raising achievements made so many hypnotic headlines.

If the Victorians put their faith in Progress, for them at least it implied a moral dimension; in these days of more sophisticated economic science all we are permitted to believe in—if 'believe' is the word—is numerical progression. Whether nation, company or individual, we have failed if we cannot show it, year by year: the logic of the system is inescapable and inclusive, and it is going to take more than a nice little phrase like 'the quality of life' to break into it.

In this force-fed economy, bent on gaining 'orbit speed' on pain of crashing, not merely over a long period, but urgently, every quarter, every year, the supreme test becomes the statistical, quantitative test. Its panicky imperatives are all-pervasive. Our cities, the heart of our culture, are now shaped by the indicators of money yield. Commenting on the shares of a famous London theatre group, Stoll-Moss, a financial editor observed: 'its chief attraction is property; many people maintain that the assets could be put to better use.' Indeed they could, by this magnificently impartial and objective criterion: the theatres could be torn down and high-yielding office towers erected on their sites. As land values rose ever higher under such 'natural' pressures, all but the wealthiest and most 'wealth'-creating users were extruded from the hearts of the cities. Tall hotels elbowed out the last residences. Urbanity gave way to nose-to-tail tourist buses; yield-per-square foot defeated the aspirations of public planners, pricing local 'democracy' out of its own domain. Meanwhile, from Moscow, the correspondent of the London *Times* complained that in seeking to create spacious city centres with gardens and theatres the Russian planners were displaying 'a cavalier approach to land values ... at odds with economic efficiency.'[14] One can only thank God that London's Hyde Park and Green Park and St James's Park, among its greatest glories, were created before 'economic efficiency' was discovered.

But such circular measures of 'efficiency'—government of numbers, by numbers and for numbers—were now widely considered to be self-sufficient. As long as statistical yield on assets employed could be shown to be rising, it was taken for granted that the public—as well as the operators involved—was automatically enriched in the waves of takeovers and mergers that ran through Britain and America in the sixties and seventies. The figures were held to demonstrate 'industrial logic', 'rationalisation', the 'economies of scale' and so on even although other results might be immense human dislocation and severe reduction in consumer choice—which was often indeed the object of the exercise.*

Nor was it any use writing to the papers to complain of the secret doing-to-death of one's favourite biscuit or beer. Market research on such occasions was subject to a sudden deafness. British breweries, once highly localised, each with its own idiosyncratic brew, adding to the wonderful variety of the English scene, had been homogenised in a great wave of take-overs and mergers and it was forecast that in this dawning age of managerial science in Britain they would shortly be reduced to six large corporations.

The only direction was Up. It was no longer permissible to cruise along on static earnings and dividend or with 'unlocked assets', however socially useful the service one might be supplying. When in the sixties the profits of the British Woolworths failed to stride forward year by year (but merely continued earning over 20 per cent on net capital), it was announced that this much deplored situation was being energetically tackled by the new chairman: 2,500 items had already been removed from the range of goods which would be further drastically pruned. More expensive types of goods were being added.[15] Yet to many, the unique service of Woolworths had always been that one could rely on finding there all those unconsidered, yet indispensable, trifles of our modern civilisation which were otherwise far to seek. However, such activities are unlikely these days to result in a high placing in the 'yield per square foot of floorspace' stakes: the imperatives of the financial Numbers Game necessitate 'trading up'.

Another valuable and in fact highly enterprising and innovat-

* An inquiry into 240 cases of takeover and merger revealed that market dominance—either by increased market share or by the elimination of competition—ranked highest among the reasons given, scoring 27 per cent of the total responses in the survey. Gerald Newboult, *Management and Merger Activity*, London, 1970.

ing British concern, J. Lyons and Co., a famous catering firm with a place in Britain's social history, fell under a similar slide-rule cloud in these years. Finally, half their unique London Corner Houses, vast, popular palazzos, and truly a part of the capital's social capital, were closed down and, in some cases, amusement arcades reigned in their stead. Again, it was not that they did not pay; merely that they did not pay *more*.

No area of life was now safe from the totalitarianism of the slide-rule. For generations Englishmen had found nothing in-compatible between good farming and the nurture of the rich heritage of the English countryside. But not any more. No economist is sterner or more 'rational' than the agricultural economist. Broiler houses, hen battery sheds, veal production units brought the computer-and-conveyor-belt approach to the once complexly organic economy of rural England. The Ministry of Agriculture reported that 500 miles of hedgerow were being grubbed up (with the aid of its improvement grants) a year; many put the figure a lot higher. However well they might serve as wind-breaks, the hedges got in the way of the big farm machines; worse, they cost man-hours to maintain. 'Every tree on my estate costs me £5 a year because of shade and roots,' complained one of the new 'managerial' farmers. When asked what his father would have thought of the 'improved' hedge-less, tree-less, and soon, probably, bird-less, landscape, he replied: 'I sometimes wonder about that, but he lived in a different era of farming.'[16]

If 'the quality of life' seemed here to have evaporated from the equation, it should be admitted that the farmer was merely applying to his industry the now universal, indispensable statistical criteria. A tree appears as a minus quantity of £5 a year because the value of the crop that could be grown in the yardage involved is precisely calculable. The spiritual refresh-ment deriving from fine trees in the landscape is not. On a wider scale, the *quantity* of food, of say, broiler flesh, is readily calculable; its *quality* or taste is not; nor are the moral and aesthetic values of animals kept in conditions of dignity, or the recreational values of the landscape in our lives.*

---

* Even so, at a later—perhaps too late—date the neglect of the facts not immediately quantifiable may make itself—quantifiably—felt. A 1971 British Ministry of Agriculture survey stated that in some areas the use of heavy

So, today, these things are subject to a deadly and systematic statistical diminution.

Statistical Diminution Systematised

Far from being restrained or counter-balanced, the system of Statistical Diminution, advancing under the irresistible banner of Science, was being rapidly pushed out to embrace greater and greater areas of life and living.

Thus, one of the great British newspaper hero-figures of the early sixties in Britain was Dr Richard Beeching, the archetypal Scientist-Manager, who left the Board of Imperial Chemical Industries to bring the full force of the Scientific Method to bear on the loss-making, tradition-rich—and ridden—structure of British Railways.

Beeching's secret weapon turned out to be a statistical survey, marrying costs and revenues over each segment of British Railways' dense 15,000 miles route network. A series of much publicised 'flow-maps' then revealed that one-third of the route mileage was carrying only about 1 per cent of the total freight and passenger traffic, measured in ton- and passenger-miles. This third rarely covered its track and signalling costs. On the other hand, half the route mileage covered its route costs six times over.

There was little very remarkable about this. A similar picture of extremes would have been revealed in surveying other public service concerns like the Post Office, electricity distribution, sewerage, education or health. But in place of the concept of service of basic social need, the Beeching inquiry located its operating criterion in 'costs-per-unit', an index capable of much greater precision. Statistical 'rationalism', rather than public reason, dictated the programme: intensive development of a few hundred miles of inter-city lines, running fast Pullman trains, with supplementary charges which would not give pause to expense-account businessmen; the 'dampening-down' of the

---

machinery and continuous cereal-growing, *aiming at ever increasing yields*, were seriously damaging the structure of the soil in some areas. Past advice to farmers, it said, had laid too much stress on immediate economics and 'present farm business management techniques should be re-examined.'[17]

heavy seasonal popular holiday traffic to the seaside since this, although it doubtless brought much enjoyment, meant maintaining idle coach capacity for much of the year; finally, the clean amputation of one-third of the railway network, 5,000 route miles, feeding the outlying limbs of the country, and some substantial areas islanded between the busiest trunk lines.

It says much for the grip purely statistical logic had now gained upon the public mind, especially when presented under scientific warranty, that these drastic proposals for the destruction of a substantial part of the nation's hard-accumulated social capital were now greeted in many quarters with an enthusiasm appropriate to the splitting of the atom.

How crisp and conclusive the equation: 1/3 of the route miles, 1 per cent of the passenger miles! In fact, it might well stand as a classic case of Statistical Diminution, or in this instance, Obliteration. One per cent of British Rail's passenger miles in 1963 would amount to 197,720,000 passenger miles, one per cent of passenger journeys to 9,384,330 journeys. Some idea of what this could mean in terms of human lives began to emerge on 19 and 20 April when the Beeching proposals were debated in the House of Commons and Member after Member stood up and related in moving detail just what a particular fraction of that one per cent would mean in terms of the life of the communities he represented.

In Lanarkshire and West Lothian, already hard-hit by the closing of the old coal pits, 'economic logic' would make the poor poorer, increasing their difficulty in travelling to jobs farther away, and discouraging new industry. For many years governments had struggled hard to avert the depopulation of the Highlands of Scotland. Now there were to be no railways —only poor single-track roads—in the five counties north of Inverness. Almost all the railway lines would be withdrawn from Wales too. Much was in difficult and mountainous terrain where the Victorian engineers had pushed railway lines at great cost and with heroic ingenuity. In such areas buses would be painfully slow and snow-bound in winter. The holiday areas of the northern West Country and much of the East Coast were likewise to be cut off from easy communication with their catchment areas in the Midlands and elsewhere, multiplying frustrations and road bottlenecks.

Although presented, as usual, as the Objective Truth, arrived at by rigorous research, the statistical indices under which all this was ordained contained, as usual, much that was in fact arbitrary. British Railways were, at many points, the victims of their long and complex history. They had had to bear the cumulative cost of land purchase, Parliamentary legislation, track and bridge maintenance, signalling systems, railway police, and until recently had been saddled with all manner of restrictive regulations. They lost traffic to road transport not so handicapped, and provided with an array of facilities, not without charge certainly (motor and fuel taxes saw to that), but with—particularly for the road-pounding heavier goods vehicles—a large degree of hidden public subsidy.

That there was a great deal of 'expert' disagreement about the extent of this subsidy* merely served to underline the spurious nature of the statistical precision which provided the explosive of Dr Beeching's demolition charge. To a considerable degree the Beeching savings were achieved by merely transferring costs—to the public authorities who would have to build improved roads, to the urban councils who would have to cope with extra congestion, to rural communities which would have to subsidise still further an already subsidised bus service. Here, as in agriculture, concentration on readily measurable yields institutionalised shortsightedness. By the 1970s, as grass grew over railway tracks, four-fifths of all the freight in this small congested island was being carried on the roads which became ever more noisy, noisome and fabulously costly as road carriers grew more and more monstrous in power and size.

It is not, of course, suggested that Britain's railway system was not in need of modernisation and some re-shaping. Electrification, dieselisation, modernisation of freight handling, the closure of obsolescent branch lines had been under way for some years before Dr Beeching's appointment. What is suggested is that in its demonstration of the capacity to 'blind by science'—statistical science—the Beeching episode marked a new high point in Britain's surrender to the Numbers Game.

* According to the economist, John Brunner, estimates of the ... proportion of motorway costs attributable to heavy lorries varied between 17½% of the cost to 70% according to whether the experts belonged to Ministry of Transport, the Road Haulage Authority, or British Railways. Article in the *Observer* 16 January 1966.

This was made clear from the outset in the curious matter of fixing the salary at which the Scientist-Manager would be willing to 'interrupt his career' to set the railways to rights. It was set at £24,000 (Dr Beeching's salary at ICI)—which was almost 2½ times the salary of his predecessor at the Transport Commission and more than three times the salaries of the permanent heads of the great departments of State. Much play was made with that figure in the headlines; the numbers of anyone who himself bore so high a number *must* be right.

The Price of Life

Having advanced its bold conclusions, the Beeching Report on 'Reshaping British Railways' nevertheless seems to have recoiled a little from its own logic. Recognising that its meticulous accountancy might have left something out, it recommended that it be complemented by 'Total Social Benefit Studies'. An economist thereupon calculated the cost of providing extra buses at £5½m–£6m, the loss of output of the men who as a result of the rail cuts would remain permanently unemployed at £1m a year, the additional cost of road accidents at £1·5m–£2m a year, the value of the time wasted by passengers having to travel on slower buses at £2·5m a year and so on.[18] When all these items had been totted up and neatly subtracted from the Beeching's operation's 'profit', the cuts were held to be vindicated since they still showed an overall, if modest, 'social profit'.

This was 'social cost-benefit analysis' which was now increasingly seen by the well-informed as the ultimate statistical key—even, some seemed to suggest, the so far undiscovered socio-economic philosopher's stone. Thus one British socialist planner, Peter Hall, described it as a departure which 'may prove to be as revolutionary in its implications as the Keynesian revolution in economics thirty years ago.' He added: 'But these studies are only just beginning. No one pretends they are perfect; they need refining.'[19]

Readers of this book may by now find the last proviso familiar, if not faintly ominous.

Like most of the statistical monitors that now shape our

lives, cost-benefit analysis originated in the United States. In the New Deal era the Federal authorities were authorised by the Flood Control Act of 1936 to participate in dam-and-river valley projects 'if the benefits to whomsoever they accrue, are in excess of the estimated costs'.[20] Since the benefits of such great schemes as the TVA were multiple and the costs were equally complex and both affected a great range of interests, cost-benefit accounting was designed to provide the proof of *net* benefit in monetary terms demanded by the Act.

By the end of the war, American cost-benefit economists had found ways of pricing the 'intangibles' and adding these into the account too; and in 1965 President Johnson vastly extended the role of the technique by carrying it from the Defence Department, where it had been used in evaluating sophisticated weaponry and immensely costly aircraft projects like the F1-11, to other spheres of government under the mystic initials PPBS —Planning, Programming and Budgeting System. PPBS, announced President Johnson, provided a means whereby national goals could be identified 'with precision and on a continuing basis' and whereby the Government could 'choose among those goals the ones that were most urgent'.

The recurring motif of the cost-benefit analysis crusaders, this notion of the superior precision of statistically determined choice, was introduced into England in the sixties by the economists imported into the Ministry of Transport by a new-broom Minister, Barbara Castle. It has been applied to motorway projects, to a new Underground Railway (the London Victoria Line) and, most extensively, expensively, and, in the upshot, notoriously, to the task of locating London's third international airport.

It is indeed revolutionary in its possibilities—but these are by no means so uniformly benign as its apostles suggest. For the basic 'innovations' of cost-benefit analysis will have a familiar ring to anyone who has travelled so far in this story of the onward march of the statistical hosts. Its distinction is that it measures both the measurable *and* the unmeasurable; and that having done so it forms them all, like and unlike, into vast, stunning aggregations of monetary numbers; and, in due course, at the end of truly heroic labours of immense ingenuity and virtually uncontrollable complexity, multiplication, addition and subtraction finally produce a neat, certified-rational numerical Answer.

Since economics is normally both defined and limited by the operation of price and the market, it follows that if cost-benefit economists are to achieve the comprehensiveness which would justify their claim to strike a true balance, they must somehow price a great diversity of things—and even non-things—which do not in fact carry prices. A great deal of their energy is therefore directed towards devising phantom markets and shadow prices, drawing complex—not to say dubious—deductions from people's carefully observed and recorded behaviour in the critical situations. Thus the research team of the Roskill Commission on the Third London Airport spent much time and money assessing and comparing the cost in noise-nuisance at each of four alternative sites. Its basic measure was the projected loss in sale value of all the houses within the plotted NNI ('Noise and Number Index') contours at each place. On the basis of a 'model' constructed from an earlier similar new airport, the 'losers' in each place were divided into three groups: (1) those who would stay and put up with the noise—who received 'noise damage' (2) those who would move to avoid the noise—who received the amount lost in property depreciation *plus* removal costs, *plus* 'Householder's Surplus' or the *subjective* valuation above the market rate of each house by its occupant; and (3) those who would have left the area anyhow—who were merely credited with depreciation.

Yet is there any real reason to suppose, that, even with a 5 per cent per annum increase built in, house price depreciation is any true measure of the full loss of 'peace and quiet' in the countryside following the arrival of a major airport? Before the testing scrutiny of the public hearings the Commission was forced to conclude: 'no satisfactory way of putting a monetary value on the nuisance caused by aircraft noise has been found.' In accordance with accepted statistical principles this did not, of course, prevent it from using an unsatisfactory one.

On the 'benefit' side of the ledger, the calculation of the sums to be credited for time saved—critical both in airport and motorway social accounting—has proved only slightly less perplexing. Cost-benefit accountants divide it into working time

saved, and leisure time saved. The first is relatively easily calculated from the normal, or overtime, wage rates of those
concerned and from the salaries of executives. But not all
will be able to agree that the high average cash figure for time
saved by in-flight business executives should weigh against,
say, low figures for time lost by ordinary mortals on the ground.
And what of the cash dimensions of time saved for leisure?
In the British M1 Motorway account it was put in at four rates
—£0.10, £0.20, £0.30 and £0.40 an hour;[21] in the Roskill Third
London Airport survey it was put between £0.11½ and
£0.34½ an hour (1968 prices) and children's time at £0.2½
an hour. But, admitted the Commission, different assumptions
about time 'led to widely divergent conclusions'. For instance,
an American cost-benefit economist, James R. Nelson, points
out that a small amount of time saved on the journey from
home to office may represent loss rather than gain since it
constitutes a 'rewarding hiatus'; he claims further that time
saved from watching, say, in-flight movies should be valued
much less than the same person's time saved at night for sleeping.[22]

High among the social benefits to be added in motorway or
freeway costing is a large reduction in the total of accidents.
Their number is fairly easily established, but to make them
addable in money terms involves putting a price on human
life itself. The cost-benefit economists do not flinch. The British
Road Research Laboratory works, predictably, on the basis of
the value of the number of years future work output lost by
the death, *minus* the value of what would have been consumed
in those years. It was then discovered that this yardstick would
result in the slaughter of old age pensioners emerging as a
clear net social gain. So a more or less arbitrary £5,000 representing the 'subjective' value of each adult life was added
to the 'objective' cash valuation.[23]*

However, not content with such reach-me-down methods,
Mr T. C. Schelling of Harvard University has elaborated a
mathematical method of arriving at the true cash value of
human life by 'intensive interview techniques' employed on

---

* The national single adult life cost used by the Ministry of Transport
(1969) was £8,800—£3,500 for future output, £300 for ambulance, police, and
property damage, £5,000 'subjective' compensation.

household groups. Men and women of various ages and classes, he suggests, might be asked 'what reduction in *income* after taxes he or she would accept in perpetuity in order to avoid a ten per cent chance of death'. Assuming this somewhat sophisticated question to have been truthfully answered, a sophisticated mathematical process could then be used to arrive at the cash value placed on say, a 1:1,000 reduction in the probability of death.[24]

It is not, Schelling reminds us, 'a particular death' with which we are concerned, but a 'statistical death'. Some of us may have difficulty in conceiving a statistical death without a particular death, but this probably only testifies to our intellectual deficiencies and inferior technique. 'Those who say that a life is beyond price,' declares the British cost-benefit economist, Mr C. D. Foster, 'do not know what they are talking about, for if the choice were between saving one life and sacrificing 5 per cent of the national income, would they argue the sacrifice should be made?'[25]

Maybe not. But the very form and blandness of this question sends a slight shiver down the spine—for it suddenly lays bare the slippery slope of the statistical 'rationality' of our time. For instance, a safety regulation introduced in Britain in 1970 required all farm tractors to be equipped with canopies to save the driver from being maimed in case they tip over—the biggest single cause of death in farm accidents. With 100,000 tractors in operation the total cost will be £40m and it has been calculated that forty lives a year will thus be saved—that is at the cost of £100,000 a life.[26]

The implication of this characteristic statistical abstraction is that this, therefore, constitutes what the economists call a defective 'allocation of scarce resources'. Statistical rationality indicates that more lives could be saved elsewhere for the same money. Quite: but *will* they be? The economists' mechanistic model lightly assumes that life follows the equations. It doesn't. Furthermore, what this overlooks is that, ultimately, this sort of action depends on generous human impulses; these, too, are limited resources. We must make the best of them when and where we have them.

Thus we arrive at the central, crippling distortion of our time, the slow, buffered poison concealed in the Numbers Game, the notion—so prevalent from the neo-Kinseyians to the psephologists—that emotion is somehow invalid, irrelevant, and even shameful. It is invalid, of course, because it defeats the meters; it is improper because the social scientist who permits himself to feel like a human being *in* society is that much less the 'scientist' objectively observing society.

This is the trap into which the Numbers Game has lured us all.

Since it is a trap both sophisticated in design and subtle in camouflage we should be grateful to the Roskill Commission on the siting of London's third international airport—probably the most comprehensive and complex exercise in cost-benefit research ever undertaken—for drawing us so excellent a diagram of it.

As the Commission had been appointed following the furore aroused by the Government's earlier (later cancelled) selection of a site, its members not unreasonably saw this new statistical technique as a Heaven-sent answer to the problem of fairly sorting out the clamorous conflict of interests inevitably stirred by so vast a project. And it is true that while—at the cost of over £1 million—the Commission did plunge into comprehensive cost-benefit analysis with aggregations of as many as twenty separate categories, as 'the best available aid to rational decision', it did also provide that 'the results must be subject to close public scrutiny and discussion'—as indeed they were for many clamorous and confusing days.

Nevertheless, dominated by economists, the Commission fell into the trap. It sought through its statistical social accounting 'to avoid subjective judgments upon any part of the problem which has aroused such deep emotion'; to minimise the area in which 'intuitive judgments are required'. It failed to see that when it came to a scheme which, basically, would obliterate five by two-and-a-half miles of close-knit English countryside, tearing apart thriving, long-established communities, and inflicting irreparable damage on an incomparable historical heritage, intuition, intensity of emotion, is as legitimate a criterion as any, and may indeed be a measure, as well as a part, of the true cost.

When the cost-benefit accountants surveyed the site at Cublington involving the destruction of four historic village churches, and the probable loss of four more, they wrote in the 'cost' of the doomed Norman church at Stewkley at £51,000. The Chairman of the County Council planning committee, however, valued this 'most splendid piece of Norman parochial architecture in Buckinghamshire', in almost perfect condition, at £5m. The £51,000 was, it turned out, the fire insurance value—most of it for the timbered ceiling and roof. On the second time round, following the local outcry, the Commission omitted the figure altogether. 'We decided,' it explained, 'that no attempt should be made to value explicitly in money terms such contentious items as the loss of wild life or churches ... We did not see any practicable way of expressing in monetary terms the regional planning factors or the value of preserving the countryside ...'[27]

With this admission, at least in a country with the close-knit character of England, the Commission's claim that comprehensive cost-benefit analysis 'seeks to bring all problems into the proper perspective' collapsed. It is a failure that bears out the comment of Admiral Rickover, a strong opponent of the wider use of cost-benefit analysis in the United States, that such procedures too often exclude the *purpose* of human effort.[28] Or, worse, beneath the meticulous statistics, the ultra-fine focus, they assume and conceal large assumptions about the ends of life which they do not state. As the United States sub-committee which examined cost-benefit analysis put it: 'It may be used as easily to rationalise a decision as to make a rational choice.'*

This is the modern magic of number, and because of its sophistication it is more subtly dangerous—more sapping—than the old. Behind the mathematical mountains of elaboration, the real issue before the Roskill Commission, as it turned out, was not really that difficult. It was whether the new airport should be constructed at high cost on empty coastal mudflats where it would do relatively little environmental damage, or whether it should be constructed, very much more cheaply, on an inland site where it would play havoc with the precious countryside remaining between the great conurbations. It was, in fact, the central issue of today's affluent societies, expertly

---

* The particular case of rationalisation they seem to have had in mind was that of the ill-fated cost-benefited choice of the F1-11, the largest military aircraft project in history.

glossed over by all major political parties: ought human life and society to be shaped to conform to 'economic logic' (that system of almost total statistical abstraction)—or should the sensitive assessment of true human needs be the starting-point, establishing the norms to whose maintenance 'economic forces' should be bent?

Despite everything done to obscure it both in figures and words, this question emerges vividly from the proceedings of Mr Justice Roskill's Commission. For this reason it may yet prove a turning-point. By leaving out, because unquantifiable, such items as 'the value of the landscape' or of the country-side, to those city-dwellers seeking recreation, this vast 2½-year-long exercise in cost-benefit accounting effectively embodies a built-in bias in favour of so-called 'economic forces'. And the Commissioners themselves bear this out. '... the future growth of the GNP,' they caution, 'must be inhibited to the extent that the choice between investments is decided in favour of those which offer lower returns.'

Of course, like President Nixon, the Commissioners wished for 'balanced growth'. It would be unfair to compare them to the old realists of Lancashire with their economic dictum: 'Where there's muck there's brass.' They acknowledge that on the basis of 'heads' there was a large majority from the start for the deserted coastal site, 'but we did not think that "simple" solution contemplated by those who had appointed us [in the Board of Trade] and devised the procedure.' So, shouldering their responsibility, the Commissioners entered what they themselves were to describe as 'the maze of Minotaur', clutching 'the Ariadne's thread' of cost-benefit analysis which would finally lead them back into the light.

There were many, reading that extraordinary Report, with its endless dedicated weighting of the perplexing detail of Time saved and Noise suffered, and the twenty other factors, who doubted whether they ever really emerged.

But one Commissioner, Colin Buchanan, Professor of Town Planning at London University, let go of Ariadne's thread at a fairly early stage. Cost-benefit analysis, he wrote in his Minority Report, 'has caused me more worry and heart-searching than anything I have previously encountered in my life.' It was not unproductive worry, however. Abandoning the slide-rules, the

ratios, the computers, Buchanan simply walked slowly through
the threatened landscapes, and set down what he felt and saw.
And as he writes, the real issue, the true conflict and the palpable weights, merely obscured by the crowding statistics
marching and counter-marching through the Majority Report,
come to life.

He writes of 'the particularly delectable open background
of London ... the softest and subtlest landscapes to be found
anywhere in the world' ... successfully preserved over the years
'more or less intact—playground for thousands, refuge for
those who seek quiet and solitude, chock-full of things to do ...'

Are these things of any value, he asks, to more than a small
minority—'perhaps somewhat old-fashioned'? And he answers
with the ring of conviction: 'Human nature does not change
so quickly that it is impossible or even difficult to distinguish
the things that successive generations commonly find of value
... the simple enduring things of life ... unsophisticated and
independent of modern technology ... and enjoyable with the
minimum of expenditure.'

His values thus 'weighed', Buchanan knew that the damage
any major inland airport would wreak would be 'an environmental disaster' and a denial of all that the great pioneer English
planners, and indeed generations of Englishmen, had stood for.
Whereas the other commissioners, following the prevailing
politico-economic orthodoxy, asserted that greater Growth (i.e.
a lower-cost airport) was essential in order to build and protect
the 'quality of life', Buchanan suggested that it was not sensible
to seek to protect the quality of life by first destroying it—
whatever the accounts might indicate.

Who's Distorting What?

It seems quite likely that had John Maynard Keynes, whose
macro-economic theory promoted the mind-stunning aggregation of the GNP, been a member of the Roskill Commission, he
would nevertheless have added his name to the Minority Report. For Keynes—in whom the economist and the artist contrived to co-exist—had himself written, more prophetically
than he knew, of 'the over-valuation of economic criteria'
bringing about 'moral decay' in our society and 'destroying the

popular ideal'. He had consoled himself by looking forward from the thirties to an age of affluence when 'we shall be able to afford to assess the money motive at its true value' and many of the 'pseudo-moral principles which have hag-ridden us for two hundred years' could be discarded.

In the meantime, to discourage over-emphasis on 'the importance of the economic problem or sacrifice to its supposed necessities [of] other matters of greater and more permanent significance', he hoped for a reduction in the overweening status of economists. 'If economists,' he wrote, 'could manage to get themselves thought of as humble, competent people, on a level with dentists, that would be splendid.'[29]

Alas, almost exactly the opposite of what Keynes hoped for has happened. Far from delimiting their function, leaving a large 'free' area where human values might develop beyond commercial pressures, economists have accepted the role of officiating high priests of our times, transforming the 'pseudo-moral principles' into something like natural law (which is less easily challenged), and now some at least march in the van of statistical imperialism, to set prices on the priceless and construct, with much ingenuity, a universal market.

Transport economists set out to handle the problem of urban congestion by 'road-pricing', levying heavy admission charges on motorists entering central areas. Education economists calculate the 'yield' on various types of 'education investment', the rate of return on masters' degrees or doctorates,* input and output and the contribution to the GNP. In the United States medical economists have been commissioned by the Department of Health, Education and Welfare to work out the cash pay-off in various programmes of medical care (e.g. lung cancer prevention versus syphilis prevention and treatment) in terms of cost per death averted, value of production in the number of years saved, and so on, determining an order of priorities following the cost-benefit ratios or notional dollar 'profit' on the operations.

It may be said that in this commercial world a realist who wishes to ensure adequate funds and status for, say, education

---

* Some recent British research, for instance, showed that masters' degrees and doctorates yielded 2·7 per cent on the money invested whereas the part-time technical Higher National Certificate yielded over 20 per cent.

or health services will demonstrate that in accountancy terms they pay off. It is a dangerously double-edged argument. While the soaring costs of medical treatment (and need they really soar so much—or is this, too, in part an aspect of the Numbers Game?) may mean that priorities must be determined, to resort to statistical 'rationality' could have insidious consequences for the ageing, the chronically sick, the 'maladjusted'—as well as for the humanity of all of us.

There are plenty of ominous signs already. 'Any man's death diminishes me,' wrote John Donne in the 17th century. More recently, explaining its cost-benefit studies, the British Road Research Laboratory observed, fairly enough, that particular cases of maimed children and broken old people who had failed to move fast enough were less likely to be effective in inducing action on road improvement than the information that road accidents in 1965 had cost the nation £650m (including £246m 'objective' costs and £305m 'subjective' costs).

The fact is—and it is a fact of which ordinary people are not unaware—that when you force prices on a good many things which do not normally carry prices, you—subtly perhaps, yet fundamentally—change their nature. Britain's Prime Minister, Mr Edward Heath, recently attempted to justify his Government's break with British tradition in imposing charges for admission to the national art collections by arguing that people would appreciate the great masterpieces more if they had to hand over hard cash to see them. On any long-term view—always assuming appreciation of the paintings themselves and not of their notional cash value is meant—almost the opposite is true: the present madly booming market in art may end by destroying all true values. These pervasive blights of our time, statistical diminution and statistical inflation, are opposite faces of the same wretched coin.

'Distortion' is the sin which economists love to denounce in others, sometimes in strangely moralistic tones, yet distortion—in a less narrow, technical sense—has been their own insistent professional offence, and as in recent years—until just now—the prestige of economists has grown and their mystique become more pervasive, so their professional narrowness of vision has spread through the whole society.

There are, however, still a few areas which remain outside

the market, real or contrived. In Britain, for instance, human blood is still felt to be beyond price. A yearly mounting roll of blood donors, reaching 1,280,000 persons in 1968, giving their blood through the National Transfusion Service without thought of payment, has more than kept pace with the soaring demand from the hospitals.

In the United States, by contrast, blood is an article of commerce, its value determined by the market. About half that used in transfusion is purchased at the going price per pint and almost all the rest secured by some *quid pro quo* arrangement—perhaps replacement of the blood by relatives and friends on a two-pints-for-one basis. 'Nothing is for nothing,' as the song says. A patient's bill for blood for a serious operation may amount to £500 and profit-making commercial blood banks are quite usual. EARN UP TO $200 PER MONTH IN YOUR SPARE TIME ran an appeal in one 'bank's' window.

It is obvious that the pricing and marketing of this unique human 'commodity' in this way—or any way—must deeply affect the way people think about it. In his book, *The Gift Relationship*, Professor Richard Titmuss has demonstrated vividly how the increasing commercialisation of blood supply in the United States has been accompanied by a steep fall in the contribution of the voluntary community donor—from 20 per cent in New York in 1956 to 1 per cent in 1966. As if to clinch matters, the Federal Trade Commission in 1966 ruled an attempt by Kansas City hospitals to use a new community blood bank instead of unsatisfactory commercial sources to be in restraint of trade, and therefore illegal.*

This is only the beginning of the social consequences that spring from putting blood into the market place. In Britain blood donors are a representative cross-section of the eligible community by class, age, sex and so on. In America, the rich buy blood, the poor sell it. It is a form of income redistribution certainly, but not perhaps the happiest that could be devised. 'Thanks to the advance of science,' reported *Medical World News* in 1963, 'the blood donors of Skid Row who at one time exchanged their blood for the price of a drink or two, can now sell it two or three times a week—at $25 a pint ... this is

---

* This verdict was overthrown by the Circuit Appeal Court in 1969, but it demonstrates the train of thought established.

threatening to turn the Bowery boys into the Bourgeoisie.'

This, in turn, leads to one more effect of market forces in their classic role of preventing 'distortion' and bringing about the 'optimum allocation of resources'. Not so unreasonably in so rawly commercial a relationship, many poor sellers fail to admit to the infections that would debar them from 'donating' their blood. The result is a high and rising toll of *serum hepatitis* in transfused patients. 'Many studies in different parts of the United States,' reports Titmuss, 'have incriminated the paid donor as the major source of infection.'

Events in Japan have followed a similar course. In the early days of transfusion, blood was given voluntarily and freely— as Buddhism at least would seem to require. But in the Japan of the 10 per cent Growth Rate 98 per cent of the blood needed is bought—in a market dominated by 'cut-price' blood banks. The rate of infection of transfused patients by hepatitis is between 10 and 25 per cent, with many deaths. Yet in spite of this harnessing of the price mechanism and the market, the shortage of blood is even more chronic and acute than in the United States.[30]

Wrong Number?

By the nineteen-seventies there were few areas where 'economic indicators' did not dominate the horizon. When Japan's fabled Growth Rate fell from an annual average of almost 11 per cent to about 8 per cent in 1971, a tremor shook the country. But the numbers in the West were in a sorrier state of disarray. A high rate of unemployment was the economist's sovereign remedy for cost-push inflation much as was bleeding for the old-style doctor confronted by any sort of fever. But now both in America and Britain high unemployment co-existed both with rising wages and a high and continuing rate of inflation. It was as if the High Priests of Ancient Egypt, reading from the secret papyruses, had announced the rise of the Nile, and the Nile had failed to rise. 'The rules of economics are not working in quite the way they used to,' observed Mr Arthur Burns, Chairman of the Federal Reserve Board, mildly enough, in July, 1971.

The trauma was the greater in that President Nixon, taking office in 1968, had proclaimed a return to the true faith in the 'natural law' of economics and had even announced an autonomous 'game plan' which would smoothly restore the self-sustaining equilibrium of the economic machine, lubricated only by the steady drip of Mr Milton Friedman's 'money supply', increasing at the rate of 2-5 per cent per annum. 'Guide lines' on wages and prices, even 'jaw-boning' the trade unions, were rejected as constituting brutal, merely human, intervention in the processes of the Market. By a singular coincidence, the Conservative Government of Mr Edward Heath, returning to power in Britain in 1970, professed an almost exactly similar faith. Adam Smith, thou shouldst have been living at this hour!

On both shores of the Atlantic, the visible hands of the official economists were thus now mainly engaged in the ritual duties of reading and announcing the Numbers put up by the workings of the Invisible Hand. But the indicators were hardly reassuring. The machine appeared to be running wild, ignorant of the 'laws' of economics. On the much-discussed 'Phillips' Curves' one was supposed to be able to read off the rate of unemployment necessary to control inflation and rising wages: about 4½ per cent in the United States, or 2½ per cent in Britain. But in both countries unemployment ran far ahead of such rates—at over 6 per cent in the US and 3½ per cent in Britain—yet galloping inflation and spiralling wages continued.

'Economists know what kinds of policy can stop inflation,' said Professor Alan Day sagely, in answer to the children's questions, 'or at least stop it from exploding. But this time we have far less evidence about the degree of intensity with which they need to be enforced ... perhaps an unemployment level approaching a million for a year or two would do the trick.'[31]

The fact was that 'other things' were now failing in their duty of being 'equal' on a scale even the purest of economists could no longer ignore. Unchastened by the presence of mounting unemployment, the big labour unions, particularly those in prosperous and vulnerable industries, were still readily able to hold to ransom the great corporations, who, in one way or another, could pass on extra labour costs to the consumer. Thus, in November 1970, after an eight-week strike which involved over a quarter-million workers in sixteen states, General Motors of America gave its workers a new 3-year contract with an estimated 30 per cent increase in wages and fringe-benefits. The following year after a nine-week strike Ford's car workers in Britain added an average 33 per cent to their pay.

Such were the rewards of the strong and well-placed. However, in the same year the British Government, powerful against the weak, broke the strike of the underpaid postmen.

But if 'other things' were now so 'unequal' as to throw even the most complexly accommodating econometric models out of kilter, after a generation of the 'touch on the tiller' view of society number-watching had become compulsive, a substitute, in the public mind at least, for any other form of action. 'Everybody is waiting for the consumer to spend the economy back into high gear,' commented the American economist, Fabian Linden, in July 1971, 'but the consumer appears to be waiting for the economy to turn round first.' Equally nervously, London fixed its eyes on Wall Street as the Dow Jones Average 'battered against the resistance level of 900 and chartists watched anxiously as it showed signs of forming a double top.'

Even so, the much signalled recovery and return to self-correcting 'equilibrium' remained elusive. In Britain inflation of over 8 per cent was accompanied by unemployment now climbing towards the million mark. In America the jobless now exceeded five million, yet in the first half of 1971 consumer prices still rose by 4 per cent.

Appearing on television at the beginning of 1971 President Nixon had declared himself, to the general astonishment, 'a Keynesian'. But even this ritual joining of the potent Visible to the magic Invisible Hand failed to remove the chronic uncertainty, the neurosis of the indicators now deepened by the anaemia of the dollar ('now the weakest major currency in the world', announced *Time*) and an adverse balance of trade.

In August, the President abruptly abandoned both the mystique and the niceties of economic science and resorted to the most direct, brutal, 'commonsense' action conceivable. To halt the inflation he imposed a 90-day wage, price and dividend freeze. To right the balance of payments deficit, he clapped a ten per cent surcharge on most imports. To rescue the battered dollar from 'the hands of international speculators' (who figure three times in the President's address although on former principles they had surely every right to be considered devoted servants of 'equilibrium') the American dollar would cease to be convertible into gold. To revive industrial investment and consumer spending, investment tax credits and other tax reliefs were to be given.

In Britain, so often now the spotted mirror-image of the United States, the Conservative Government also was now being forced away from its initial position of non-intervention in the workings of Economic Providence. A 'Keynesian' reflation through purchase-tax removals was to be accompanied by a 'voluntary' Confederation of British Industry-organised price-freeze.

The President, announcing his 'new economic policy', appealed to the patriotism of Americans. The British Prime Minister, Chancellor and the Minister of Labour appealed for a sense of responsibility, particularly by the workers. But in an economist-dominated world, patriotism is hampered by the profound disadvantage of having no price-tag, and 'responsibility' inevitably evokes questions such as 'to whom' and 'for what'.

To the GNP? Or to the accountants of 'yield'—per £ or $ invested, per man, per square foot of ground? To some 'whizz-kid' master of the Numbers Game, building a quick fortune and financial empire by shrewd opportunist take-over and asset-stripping? To a society and government whose only criterion of the equity of pay awards, and rewards in general, turned, in practice, on the brute bargaining power the claimants commanded or the vulnerability of their employers or the industry's ability to absorb costs, perhaps by transferring them to the consumer? (Even 'productivity' agreements which have almost monopolised any argument from 'equity' are, in fact, necessarily highly capricious in their incidence and derive from accountancy rather than social equity.)

Social responsibility derives from and begets social responsibility, and there can be little of it in a society which takes its standards not from the forum, but from the market, not from the community, but from a semi-automated, self-validating Numbers Game such as encloses us today.

On both sides of the Atlantic many of the most experienced economists were now insistently calling for that new form of Visible Handery called an 'Incomes Policy'—prominent among them a former dedicated apostle of the sovereignty of the market, the chairman of the Federal Reserve Board, Mr Arthur Burns.

Thus, a year after the Conservative Government in Britain had abolished its predecessor's Prices and Incomes Board and rejected an overt Incomes Policy, in the United States, in November 1971, President Nixon boldly embraced one, setting up a tripartite Pay Board, a Prices Commission, a Productivity Commission and a complicated apparatus of wage, price and profit mandatory guidance or 'yardsticks'.

Referring to the negotiation of 'a sort of social compact', *Time* considered that 'the President's actions have changed the American economy for years to come, perhaps forever.' But what was offered—avowedly as a stopgap until the free market mechanism could somehow be repaired—was, in the main, one more complex statistical system with the statistical goal of slowing down the rate of inflation to between 2 and 3 per cent by the end of 1972. The President's appeal was to patriotism but, as with the earlier British 'Incomes Policy', the substance was a sterile managerialism, statistical elaboration, the print-out of the computers.

It is not of such stuff that 'social compacts' are made. Predictably, President Nixon called for the 'participation of all areas of society'; predictably, labour leader George Meany, like his British opposite numbers, declared that the Pay Board was a 'stacked deck' against the workers. For the endless, compulsive struggle for more pay is not, at bottom, a technical question of 'productivity' and 'differentials' and so on, however much these are debated, but the basic political conflict over the splitting of the national income. It has to do, not with statistical mechanisms and adjustments, but with the human need for a just society, and for social justice to be done and to be seen to be done. (If, however, grab is to be the only visible rule, we shall all equally need to demonstrate our manhood by grabbing.)

In a dawning awareness of this, some of those who had put all their trust in the omnicompetence of the economic mechanism and its regulators were now looking to politics to put

Humpty-Dumpty together again. But it was late—perhaps too late. For the economists and other exponents of statistical automation had succeeded all too well, in some respects at least.

Index economics had given birth to index politics.

WHAT PRICE DEMOCRACY?

There was no issue when it came to selling Ford automobiles; there was only the product, the competition, and the advertising. He saw no reason why politics should be any different.

> Joe McGinniss, *The Selling of the President*, quoting Harry Treleavan, communications executive in the Nixon 1968 Presidential election campaign.

Those who live by numbers can also perish by them and it is a terrifying thing to have an adding machine write an epitaph, either way.

> 'Adam Smith', *The Money Game*, 1964.

There had never been an election in Britain like the General Election of June 1970, and one may hope—almost certainly vainly—that there never will be again. It seemed to unroll like a print-out from the semi-automated computers of the five opinion-polling organisations, and if the ultimate conclusion did not quite coincide with the scenario that too seems, in retrospect, hardly more consequential than a sudden change of luck at Bingo.

With the lowest turnout (71.5 per cent) of any general election in Britain for a generation, the electors remained from the first shadowy figures, of importance mainly as the raw material of the pollsters and psephologists. It was an election dominated not by people or politicians, but by Numbers. It was numbers, numbers, all the way.

By way of confirming the Labour Party's historic mission, or the claim to have 'Life and Soul' as the smartened-up slogan had it, Mr Harold Wilson centred the campaign almost wholly on his Government's crowning glory, the achievement of a £550m balance of payments surplus, 'the largest we have ever had'. On his side, the Conservative leader, Mr Edward Heath, like the principal of a rival firm of accountants, endlessly disputed his opponent's figures, and drew the nation's attention to the price of jam. If Labour came to power 'we would have a 3-shilling loaf, a shilling bus-fare and a shilling telephone call, and we would have to pay for them out of a 10s pound.' In Leicester, picking up the mantle of Disraeli, he evoked the spectre of 'milk up twopence a pint, jam up eightpence a pound, sausage up ninepence, coal up £2.10 a ton.'

This was the Conservatives' great 'Housewife's Shopping Basket' theme, said to have been devised by the campaign consultant, Mr Geoffrey Tucker, a former advertising account executive, psychologist and statistician—that potent recurring combination. In addition, the Party's Research Department had looked up the previous year's newspapers to pinpoint the 'economic indicators' likely to emerge before Polling Day.[1] With the help of these they planned to reactivate the neuroses nourished by two decades of economic numerology, creating—despite Harold Wilson's boasted 'national surplus'—an atmosphere of economic crisis.

Yet the mock battle of the economic indices formed merely the counterpoint to the grand daily—indeed it seemed almost hourly—alternation of the rival opinion polls whose decimal-point divergences achieved a 'reality' which eluded the contenders themselves—even when some much-headlined 'gain' fell within the sampling margin of error, and might be wholly bogus.

The newspapers seemed to set much store by the prestige-value and pulling-power of the polls, combining as they did the best features of the astrologer's columns and the racing odds plus the added bonus of respectability as being 'scientific research'. They ran them for all they were worth. No fewer than 30 per cent of the lead stories in the major national dailies and Sundays in the three weeks before election day were about the verdicts of the polls, and even on those days when they did not make the lead, the pollsters' findings figure prominently.[2] Television too was equally convinced that 'Polls Make News'.

And not only 'news'. Since April, when the polls showed the strong Tory lead of the New Year to be slipping away, the election date seemed to have been writing itself on a statistical planchette. By 12 May Gallup was recording a 7½ per cent

Labour lead; a day or two later all five polling organisations had lined up, giving Labour an average lead of 3.4 per cent. The Labour Government's term had still many months—theoretically almost a year—to run. It was, in fact, in the middle of a critical series of economic manoeuvres. But whether an election was in the national interests was scarcely discussed. Blazoned in the headlines, urgent in the charts, the polls' figures seemed to carry their own imperative. Announcing a June election, the Prime Minister explained that 'an electioneering atmosphere was developing ... Everyone seemed to want an election.'

Everyone? Want? But the same curious numerical automation persisted. The day before Wilson went to see the Queen, *The Times* came out with its new poll: MARPLAN GIVES WILSON 2.7 PER CENT LEAD EQUAL TO 60 OVERALL MAJORITY. The statistical drama intensified; in the three weeks of the campaign the five poll organisations between them published twenty-five sets of results, while the *Sunday Times* stole a march in numerical necromancy with its 'Poll of Polls', a 'scientifically consolidated' index. Two weeks before election day the veteran political editor of the London *Evening Standard* explained across a full page 'Why I think the Election's Already Over: For the first time all five teams of pollsters are clear and unanimous in their prediction.'

This abdication of journalistic judgment and legwork before 'science' was now widespread. Having, it seems, become converted to Professor Lazarsfeld's view (Chapter 6) that social life had become 'far too complicated to be perceived by direct observation', editors were hiring their own university psephologists to perceive it indirectly, by endless analyses of the analyses of the opinion polls. These largely eclipsed old-hat reporters in the field as with the aid of intersecting graph-lines and multiple correlations they explained why Labour would win and, then, with equal authority and even more refined graphs and correlations, why Labour had, in fact, lost.

Under these auspices, as everybody watched the indicators, concentration was on technique rather than content. Would Wilson's 'low-key' approach pay off? Did the latest poll figures suggest that Heath had made a fatal mistake in leaving his economic 'counter-offensive' so late? Who was hogging most TV attention? Whom did the 'ratings' show coming over 'most

sincerely' on the small screen? And so on, drearily, ad infini-
tum. 'We virtually didn't raise any issue,' admitted the Labour
Minister, Richard Crossman, afterwards. The election, Lord
Watkinson, a former Conservative Minister now chairman of
Cadbury-Schweppes, explained in a letter to *The Times*, was
about rival 'management styles'.

By the time Polling Day arrived the actual intervention of the
voters had begun to appear almost superfluous. In the event, it
wasn't. Yet it would be rash to hail this as evidence of the continuing vitality of democracy, because the probability is that
what defeated the phoney precision of the pollsters' monitoring* was the dramatic use of 'economic indicators' even more
hollow. On the last Monday before polling day the Tories'
eagle-eyed watch on these potent numerical 'signs' was abund-
antly rewarded by the appearance of the routine May
balance of payments figure which showed a deficit of £31m.
Like any single month's figure, it was virtually meaningless—
and, as it happened, swollen by a single sum of £18.5m paid
for two Boeing 747s. But it was enough. 'The Labour case,'
proclaimed Mr Heath firmly, 'has now collapsed.' Devaluation,
it was hinted, might be the next step.

The same day the index of industrial production figure for
April was released. It had fallen—to 125 from 126 in March.
The fall, pronounced Mr Iain Macleod, the Tory Shadow Chancellor, was 'alarming when taken in conjunction with the
recent trade figures'. Industrial production, he explained, was
now running nearly ten per cent below the level it would have
reached had the growth-rate of the last 5½ years of the Conservative Government [to 1964] been maintained.'

It was truly an awesome tribute to the pathetic state of
numerical neurosis or statistical jitters to which the British
people had now been reduced that this sort of statistical
shadow-play could be taken seriously and could succeed. But
nowhere did the picture of a nation besotted with the Numbers
Game come over more clearly than in the grand electronic
finale—the now institutionalised TV circus of Election Night
with its bizarre blend of old political ritual, modern showbiz,

* All the polls save one ended with figures belied by the result (if one
disregards, as ever, the standard margins of error)—an overall Conservative
majority of 31.

and up-to-the-minute pseudo-science.

Both BBC and commercial channels ushered in what must surely now be called 'The Mother of Parliaments Election Show' with a gaggle of well-known entertainers who filled the hour before the first result came in with what one TV critic characterised as 'eruption of gags, sneers, insults and cracks against all politicians'. This, however, was merely the curtain-raiser before the entrance of the real star of the show, the Super Computer from which, one was given to understand, Britain's political destiny would shortly emerge. It was registered, from moment to moment, on an immense dial with pointer, indicating 'The Swing'—which the viewer might have been forgiven for taking for some sort of natural force, akin to gravity. 'BBC Election Researchers at Nuffield College, Oxford,' had, it was explained, 'for the first time computed the social and economic factors for every constituency in Great Britain,' and when the early results were fed into the computer, this would enable the attendant team of political experts, psephologists and statisticians to 'predict the outcome for each of the results to come' and for the entire election.

The historian of the future will surely find some fascinating footnotes here. He may speculate, for instance, upon the mysterious compulsions that drove a patient, politically mature people like the British, after weeks and months of almost continuous political prediction, to go to such immense pains and expense to anticipate, very faultily, with many hasty corrections, by a few hours, election results which they were in any case now powerless to change.

But it is doubtful whether he will get many clues from the BBC itself which, after having resisted the Numbers Game so long and so stoutly, was now among the most accomplished players. As an epilogue to the show, it announced that while the early results were being given out, 15m viewers had been tuned to the BBC against 4m to the commercial stations; and even at 4 a.m. 600,000 still stuck with the BBC compared with a mere 200,000 with the other establishment.

'Scientific Democracy'?

But the British General Election of June 1970 marked much more than the supersession of the traditional Westminster five-yearly reckoning by the Princeton 'continuous audit'. It registered a new stage in the progressive debilitation of political life by the convergence upon it of the various aspects and inter-locking devices of the Numbers Game, which was finally to leave man trapped within an automated politico-economic process he seemed to lack both the power and the will to break out of.

As ever, America blazed the trail, but the British were proving more and more adept at picking it up. Thus, in January, 1969, with a flourish of trumpets, the London *Times* introduced its 'new and more refined' Marplan leadership rating index (making possible measurement of 'the leadership qualities that influence votes') which was, in fact, a modification of the 'Semantic Differential Test' employed in President Nixon's campaign of the year before.

Described as 'the most sensitive instrument known to modern marketing research',[3] the Semantic Differential Test employed in the Nixon campaign consisted of twenty-six pairs of anti-thetical adjectives—tense relaxed; stingy generous; cold warm; stuffed-shirt sense of humour, and so forth—strung out on a seven-point scale of gradation from one pole to its opposite. In the 1968 campaign a social scientist commissioned by the Nixon management had travelled through the country asking people to evaluate the presidential candidates on each level of the scale and also to indicate upon it the precise mix of qualities each felt the Ideal President should possess. By calculating the average desired score for each quality, and linking these with a line, the Ideal President Curve could then be plotted. Comparison with the curves that emerged from the evaluations of the actual, existing, un-ideal candidates then exposed 'the Personality Gap', while at the same time pinpointing Nixon's particular weaknesses *vis-à-vis* the Ideal in comparison with his rivals. And this, together with other charts breaking down the response geographically and ethnically, suggested the most necessary 'image' corrections.

Coupled with the advent of the professional campaign consultant and attendant social scientists, such sophisticated devices, mark, according to James M. Perry, a senior editor of *The National Observer*, the arrival of 'the new scientific politics'. In its 1966 Congressional campaign, the Romney Organisation 'used all kinds of specialised techniques' and 'each was carefully market-tested'. In his 1967 campaign in the New York State Governorship election, with its 3,000 commercials, Nelson Rockefeller equally showed himself a master of 'the new politics'; while 'Romney had been able to recruit young professors, Rockefeller hired deans.'[4] According to the pollster Louis Harris, 80 per cent of American Senators are now commissioning private polls to assist them in tailoring their campaign appeals. With two hundred private pollster concerns at work measuring the grass-roots, political market research appears to be almost as flourishing in the United States as its commercial equivalent.

Perry claims that the new 'scientific' approach makes for greater political flexibility. This may well be true of some fossilised political conditions. It is true too that modern democratic politics are, to a degree, inevitably of the market-place. Ever since Queen Victoria complained so bitterly of Gladstone's '*constant* speeches at every station' in his great Midlothian campaign of 1879, even in Britain a political leader has been obliged to 'sell himself' and his ideas to the mass electorate.

Even so, there is a vital distinction between the flexibility that responds to the specifications of a mechanical or statistical monitor, and the intuitive flexibility of the true political leader. The one is a deadening and demeaning process; the other is alive and may be richly creative. Can one imagine Lloyd George's ringing Limehouse onslaught on ducal landed proprietors being assembled in cold blood to correspond to the data of an attitude survey? Or, for that matter, Theodore Roosevelt's summons to the strenuous life—or Franklin Delano Roosevelt's heart-warming confidence in the days of despair, taking form in response to the Ideal President Curve?

These multiplying mechanistic devices find their moral sanction in the Great Democratic Fallacy, as American as blueberry pie, but now exported in many glossy packages, that it is possible to synthesise 'What the People Want' by totalling the items ticked by constituent individuals. Since Bryce warned against the danger of excessive faith in the omniscience of the greatest number, statistical techniques have renewed the mystique. The 'dynamic' mathematical model has now made its appearance in American politics. For the 1960 Kennedy Presidential campaign a group of businessmen and social scientists set up the Simulmatics Corporation which built a 'simulator' and stocked it with data from sixty-six earlier opinion polls held between 1952 and 1958. American voters were divided into 480 types (by combinations of location, income, ethnic origin, religion and so on), and each group linked to the attitudes it had previously displayed on fifty pre-selected issues. Fed with alternative lines of appeal, the model was supposed to deliver a complete numerical summation of all reactions. However, writing in 1964, the simulator's designers did point out that it was still 'rudimentary'.[4] With the faster computers coming along, they said, much more might be expected.

'Somehow False'

In Britain in the mid-fifties, only lately introduced to the television commercial and the natural break, there was still some resistance to political marketing. The Labour Party, stemming from a democratic movement which historically was fired by a sense of mission, shrank from the notion of hiring mercenaries to tailor its Message. As late as 1957, despite the expensive campaigns initiated by Lord Woolton for the Conservative party and beamed at the younger, better-off worker, Mr Hugh Gaitskell was still voicing what was still a widely felt distaste for professional advertising in politics—'the whole thing is somehow false.'[5]*

Nevertheless a year earlier, as party leader, Gaitskell had

* An impression possibly strengthened in Britain by the battery of advertising techniques used to sell General Eisenhower as President in 1952. Cf. David Ogilvy: 'In 1952 my old friend Rosser Reeves advertised General Eisenhower as if he were a tube of toothpaste' (*Confessions of an Advertising Man*). However, commercial agencies had been used in US Presidential elections since 1936, when the Republicans sought to dramatise the issues by hiring an advertising agency famed for its radio soap operas.

already commissioned two surveys from the Mark Abrams market research firm, and in 1960 the Gaitskellite *Socialist Commentary* commissioned another, designed to reveal 'why a party like the Labour Party has been losing appeal. What are people's aspirations nowadays? What do they hope politicians can do for them?'[6]

Dr Abrams detected a swing away from 'ideology' and towards 'consumerism', and at the next election, in 1964, was engaged by the party to delineate the 'target voter'—the voter who would have to be 'sold' to win. This is a concept derived direct from marketing and advertising, and it was now merely a logical extension that a group of professional advertising men —volunteers rather than mercenaries—should be harnessed to the indicated task of getting rid of 'the old puritan image' and substituting 'something dynamic'—like (an early inspiration) the slogan LET'S GO WITH LABOUR and 'thumbs up' sign.

Earlier notions about the 'falsity of commercial advertising' having now apparently been discovered to be old-hat, the party's professional 'communicators' devoted eighteen-months' research to image-building before the 1970 election. By 1969 they were able to announce that their researches had revealed that Labour should be 'projected as a party of idealists'[7] in a campaign with the key slogan 'Labour's Got Life and Soul'.* To ensure that the point was taken they provided a variety of stickers and badges such as 'Get Life and Soul and Win', 'I'm a Soul-mate, Mate', or, simply and spiritually, 'SOUL'.

However, possibly concluding that a party actually alive would not need to harp on the fact quite so much, the electorate in 1970 apparently found more persuasive and target-worthy the other side's line, 'The £ in your Pocket is Now Worth 15/10—You'd be Better-Off Conservative,' also the brain-child of an advertising man, Mr Geoffrey Tucker, formerly of the US Agency, Young and Rubicam, also of Colman, Prentis and Varley.

---

* According to the *Sunday Times*, the Creative Director who originated the slogan 'came to the notice of Transport House' via campaigns for Danish bacon, Durex contraceptives, Aspro and non-woven disposable pants for women.

The Disintegration of Politics

By 1970, in Britain as well as in America, politics could be—and indeed was—resolved into a seismic pattern of intersecting graph-lines not greatly different from the charts of the stock market or television audience ratings. In Britain particularly, long famed for political stability, the measured alternation of the great parties had been replaced, on paper at least, by abrupt swings of allegiance, sometimes from month to month. In October 1965 a baffled Conservative MP wrote to *The Times* to point out that in early August 'the polls' had indicated a Conservative majority of about 150 and in early September a *Labour* majority of about 150. (He hadn't, he said, noticed any such change when out and about in his own constituency.)[8]

Replying to the puzzled MP, Dr Henry Durrant, then managing director of the Gallup Poll in Britain, explained that 'this new behaviour is now world-wide ...' It was what the social scientists had dignified by the name of 'Volatility'. Dr Durrant put it all down to the disruptive 'impact of technology'.[9] But he modestly refrained from including under that head the continuous electronic pulse-taking of the polls themselves. Indeed, any hint that this might have something to do with it was liable to be met by pollsters and allied social scientists with a mixture of indignation and pitying contempt for lay naïveté.

And yet it was hard to see why unrelenting measurement of every quiver of the public mood, every transient reaction to the headlines, should not give rise to a certain volatility of 'opinion'; why what a political correspondent described in 1964—as he might have in 1970 and other years—as 'another perplexing bumpy ride on the opinion poll seesaw' would not promote a certain disorientation. Dr Durrant himself in his letter to *The Times* pointed out that 'volatility' was particularly high *between* general elections—i.e. when people's responses do not 'count' and may be lightly thrown out.

What we have here, as already suggested, is the 'counterfeit' presented as the real—another addition to the statistical phantoms and pseudo-events that clutter our times. Does it matter? The greater fluctuations, Dr Durrant pointed out, were accompanied by a higher percentage of Don't Knows than ever before, and in a world where it becomes more and more difficult to distinguish between the bogus abstraction and the real there was much to suggest that this was in fact the statistical face of 'Don't-Want-to-Knowery'. The 71.5 per cent turnout at the British General Election of 1970 represented not only the lowest figure since the mid-thirties, but a continuing decline at every election save one since 1950. In America and Britain alike the young voter was reported to be notably apathetic.

To extend the brittle 'racing odds' approach of election times by publishing continuing 'odds' over half the term of a government must tend to substitute ticker-tape price-making for genuine dialogue and the development of true opinion. In fact the plethora of public and private opinion polling, of market research, of psephological analysis, the whole statistical bag of tricks, makes not only, as usual, for the eclipse of the non-quantifiable, but for the extrusion of live issues and creative controversy altogether. For Dr Gallup and other pollsters agree that when presented with a range of alternatives, respondents to mass opinion polls tend to cluster around the middle. This *may* be due to studied moderation. It may also be due to indifference or ignorance, seeking the obvious 'out'.

Since then this sort of phoney 'consensus'—the consensus of sustained indifference—has spread widely. In both the 1966 and 1970 British general elections and the Greater London Council elections the Conservatives soft-pedalled the 'comprehensive' school issue, although selective, middle class, 'competitive' education is near to the heart of their creed.[10] Similarly, the Labour Party omitted from its 1970 election manifesto any mention of a wealth tax, proposed in a report to its National Executive not many months before, although again, one would suppose, a fundamental of a socialist approach in a country where 2 per cent of the people continued to own almost 30 per cent of the wealth and 25 per cent to own 90 per cent.*

In Britain as in America the parties now assiduously sought to avoid true debate at elections, preferring controlled interviews and market-researched and packaged presentations to confrontations on television, or even to appearance before live, unrehearsed audiences. Leading his party into 'battle' in 1970, Mr Harold Wilson cried: 'The public does not want a lot of

* *Social Trends* No. 2, 1971, Government Statistical Service.

argy-bargy ...' What the public wanted was 'quiet, strong government'.[11] The Conservatives, it seemed, agreed. They presented on TV 'a young housewife' named Sylvia, deciding to 'give them a trial'.

In his study of the 1970 British Election, the psephologist, Dr David Butler, accuses the electorate of 'tempting' politicians to focus so heavily on 'short-term economic benefits' since they, 'the electorate did seem to see politics increasingly in bread-and-butter terms.'[12] But, in fact, for years the public opinion polls have been nagging the public for their views on 'the standard of living' and whether it is going 'up' or 'down'. Like Dr Durrant's this is a case of that not unimportant phenomenon, social scientists' invisibility. They are so professionally focused on their subjects that they cannot include the effects of their own public studies and diagnostic devices. The light strikes them, but passes through.

The 'Credibility Gap' and All That

The devaluation of politics was accompanied and speeded by the devaluation of politicians, who no longer received the full-bodied denunciation that stems from genuine feeling, but were merely accorded the 'marks' of the condescending school report. Although a survey showed that almost a third of the British public could not so much as *name* five leaders of any of the political parties, this now need be no bar at all to endorsing some such prefabricated verdict as 'could do better', or 'I dislike him in some ways, but he does have some good points'; and the pollster organisations and their newspaper patrons could be relied upon to issue frequent succinct summations of these scientific versions of the 'Thumbs Up' or 'Thumbs Down' of the Roman Arena. Thus the National Opinion Poll of February 1968, on the then Conservative Opposition leader: 'Mr Heath's trouble is that he is widely felt to be a weak leader, and out of touch with the general public. In addition, he faces the problem that two-thirds of the electorate prefer their Prime Minister to be married ...'

Still lagging behind as ever, Britain had not yet achieved the precise calibration of the 'credibility gap' in her political leaders attained in America (in April 1971 the Gallup Poll showed that 69 per cent did not believe Nixon on Vietnam compared with 65 per cent who did not believe Johnson in his first full term), but the Recording Angels of BBC Television would sternly call the British politicians to task over their ratings. Thus Robin Day to Edward Heath (when in opposition): 'How low does your personal rating among your own supporters have to go before you consider yourself a liability to the party you lead?'[13] A question with a certain impact, because in July 1965 the National Opinion Poll had found the then party leader, Sir Alec Douglas-Home 'less sincere' than Mr Harold Wilson, and a week later the ex-14th Earl had resigned the leadership. (And while according to Richard Hodder-Williams* 'it would be wrong to imagine that there was a direct causal connection' between these events, the ratings nevertheless exercised an insidious, creeping blackmail.)

Odd to reflect that only five years ago the Conservative Party had not even elected its leaders. But then that was aeons ago, before politics had joined the great world of showbiz and before Marplan's leader index† had come to the aid of a gentleman's intuition in these things. Now reasoned, straight political broadcasts are felt to be a bore, an intolerable overhang of the 'square' Reith era, and the political leader who wishes to do something for his rating appears on the Frost programme or sings a duet with Ena Sharples in the soap opera, 'Coronation Street'.

Unfortunately, at this point it is apt to be the professionals who command the stage, the politicians who become the 'feeds'. A Frost or a Jack Benny outlasts a Macmillan, even with the white fur cap, an LBJ even with the 'Great Society'; and now it is suggested that a Prime Minister of Great Britain whose rating is drooping should be taken out of the show. In 1969, a leading British psephologist, Professor R. T. Mackenzie, asked

---

* In British Opinion Polls and British Politics, London, 1971.

† Cf. Derek Radford, director of Marplan, in a letter to The Times: 'The image which a person creates—in whatever walk of life—is indeed a highly complex matter, but fortunately interviewing and statistical techniques are available which make it possible to identify and summarise the many and varied qualities which go to make up the total image of a person, and enable them to be summarised in simple arithmetic form—or rating—which can be readily understood.'[14]

Harold Wilson in a TV interview whether he did not 'reach a stage where his appearance has become over-familiar, not to say boring'[15] and in a memorandum to party colleagues about the same time, Commander R. T. Paget, QC, MP, suggested that such was the contemporary 'level of exposure' that Prime Ministers must be considered expendable after a maximum of 'two or three years', after which 'the credibility gap' tends to become excessive[16]—and also, presumably, the entertainment value is exhausted.

The only doubt must be how long any replacements will be available. For like monetary inflation, political devaluation is progressive and self-feeding.

'In order to test a hypothesis ... that the public is now taking a thoroughly jaundiced view of politicians in general,' the National Opinion Poll early in 1968 put to the British public, for assent or dissent, nine propositions about its politicians. The poll claimed they were all 'based on comments made by electors in recent NOP surveys'. Be that as it may, they were all negative.

It was duly announced that 78 per cent of British voters felt that 'most politicians will promise anything to get votes'; that 57 per cent 'agreed' that most were 'in it for what they could get out of it'; that 66 per cent believed 'most cared more about their party than their country' and so on.

In fact, it is perfectly possible for 'most' people to assent to such worldly-wise sentiments on one level while remaining aware at a deeper, less saloon-bar level that, in Britain at least, whatever may happen elsewhere, many 'politicians' are dedicated individuals and willing workhorses of democracy who deserve our respect and gratitude. Yet with the fixatives of print and number the cynical mood may be installed in permanence. The danger is great: for, more than ever before in history, our 'advanced industrial society' is formed in the mould of statistical Expectation.

Words and Numbers

The erosion of the old political values—and substance— proceeded simultaneously on two fronts. As the outer was

taken over by the psephological monitors and scientific salesmen, the inner was occupied by the economists with their
equally bleak furniture. The combined effect on the parties of
the left, from which the dynamic of political and social
evolution would normally have derived, was peculiarly blight-
ing.

Succeeding to the leadership of the British Labour Party in
1955, Hugh Gaitskell solemnly warned fellow members against
'the subtle terrorism of words'—by which he meant continuing
faith in those time-honoured slogans such as 'common ownership', which carried almost religious overtones, verbal symbols
of human aspirations akin to those represented by the word
'Liberty' among those truths the Founding Fathers—'held to be
self-evident'.

Had Gaitskell just then spoken of the 'subtle terrorism of
Numbers' he would have been on firmer ground. For Numbers,
not words, were on all sides in the ascendant. In his book on
the statistical projection of the Future, *The Art of Conjecture*,
the distinguished French economist, Bertrand de Jouvenal, sees
the main function of politics as the securing of that stability
which will enable the economists and other social scientists
to get on with their forward planning. There is little to suggest
that Britain's new generation of Socialist Prime Ministers
would, in their hearts, have dissented from such a view. For
they were themselves professional economists. And while,
unlike the Wykehamist Gaitskell, Harold Wilson pronounced
'disequilibrium', 'productivity' and 'gross national product'
with a Yorkshire accent while puffing the pipe of class 'soli-
darity', his preoccupations in power were equally those of the
trained economist and Fellow of the Royal Statistical Society,
and his style—as the 1970 election was to show—was very
much the modern 'scientific management' style.

In 1956, a year after Gaitskell's accession, another Labour
economist, and future Minister, Anthony Crosland, hastened
to follow his leader's behest and strike off the shackles of
tyrannical words in an accomplished exercise in revisionism,
The Future of Socialism, in which earlier generations of the
faithful were magisterially rebuked for cherishing 'the vulgar
fallacy that some ideal society can be said to exist of which
blueprints can be drawn'. Whereas such economic illiterates

had seen 'the System' as the source of the distortion of human nature and society, it now appeared as a mechanism of great intricacy and power (and of course the 'given' of the economist) which human blundering was in danger of damaging, when in fact any malfunctions could now be readily corrected by economic expertise without impairing 'organic unity'.

Like President Nixon calling for 'balanced growth', the new British socialist leaders, completely freed at last from the 'tyranny of words', now aspired to civilise capitalism. The characteristic initiatives of the British Labour Government under Harold Wilson were directed towards making capitalism more 'efficient' through such institutions as the Industrial Re-organisation Corporation, in effect a publicly financed merchant bank, run by ex-merchant bankers, dedicated to promoting take-overs and mergers in line with 'industrial logic'.*

Only a few years earlier British Labour governments had placed the ideal of 'public service', and the public service organisation boasting wider aims than the maximisation of financial yield, in the centre of its perspective. Now, with the zeal of the new convert, the Labour Government set about turning even Her Majesty's Mails—the Post Office—the classic instance of a public-service activity—into a commercial business corporation, whose surpassing merit was that it would respond to financial criteria. It did so: soon postal charges were being repeatedly increased, service offered was being drastically cut down, and, at a time of high unemployment, the number of postmen was being reduced.

For the benefit of the less numerate members of the Labour Party for whom the progression from Blake's Jerusalem to the new statistical goals might appear disturbingly abrupt, it was customary to follow the now central themes—'Can Labour Achieve a High and Sustained Growth Rate?' with a parenthetic reference to the need to finance progressive social policies, hospitals, schools and so on where appetite for cash was insatiable. Yet somehow these 'ends' had a way of receding

* Cf. the First Annual Report of the IRC on the value of 'IRC's unusual status, created by Government, but directed by businessmen, free to make its own decisions about individual projects without political bias and not subject to Government veto.'

into the distance while the 'means', in all their meanness, eclipsed them.

For the ends were words and the means were numbers, and in the economists' society numbers prevail. One might suppose, for instance, that the 'Incomes Policy' of a socialist government might have something to do with the effort to attain social justice or at least a slightly less flagrantly inequitable distribution of wealth and income. In fact, it was merely a statistical mechanism, geared to statistical criteria (of dubious content) like 'productivity' and to the maintenance of existing differentials. In October 1970, Sir Jack Scamp, chairman of an independent inquiry into a protracted strike by low-paid municipal sewage workers, dustmen and others, pointed to the need for a national incomes policy which 'can succeed in restraining increases for high-paid workers while the lower-paid improve their relative position.'[17] No such policy existed, his report concluded.

Nor did it appear very likely to, for both the creative vision to act outside the economic mechanism and the moral vision to support values other than price now appeared almost totally eroded. Numbers had triumphed, and there remained nothing but the sovereign 'non-ideological' ideology of economics, with its grotesquely stunted view of human motivations, founded long ago on metaphysical and mathematical abstraction, rather than on detailed observation of people in action.*

Economically, man's wants are insatiable, morally equal, interchangeable on the basis of price. If economics takes over the role of politics, so does this view. Disintegration is progressive. The ideal of public service was once a reality on all sides of politics in Britain. It was one of the country's great strengths. But it is not really compatible with a system which evaluates men according to the salaries attached to their names or with mechanical insistence on the highest market 'rate for the job'. Recently the Vice Chancellor of a British university was quoted as complaining: 'I'm really chairman of a company

---

*Thus Professor J. K. Galbraith ranks as nonconformist, if not revolutionary, when in The New Industrial State he suggests two additional spurs, besides the wish to consume and the fear of hunger (the celebrated 'stick and carrot'), particularly since both were non-material.

with a turnover of £10 millions, dealing in knowledge in penny packets; and they argue about paying me £7,500 a year.'[18]

For what is that so characteristically contemporary phrase, 'the rate for the job', but one more self-sufficient Number, blandly begging all the real questions, smartly rounding out yet another circle, smoothly lowering the curtain on all consideration of values rather than prices, and on troublesome words such as 'social justice?'

Having brought Britain to the brink of industrial breakdown and disaster in the miners' strike of February 1972, Mr Joe Gormley, President of the National Union of Mineworkers, observed: 'It is a hell of a thing in a civilised society when you have to use tactics like this to get the wages the miners are entitled to.'

It is indeed. Yet in this society of the Numbers Game there was no alternative in sight between crude force and statistical sterility, between our bizarre, half-stylised big Union—big Corporation confrontations, reminiscent of the baronial jousts of the Middle Ages, and the dehumanised, mechanical and often meaningless pedantry of 'differentials' and productivity indices and the clever point-winning ball games that go by the name of 'Incomes Policy'.

Pursued over long periods, with no escape in sight, both alternatives took their moral toll. Swinging between them was perhaps the most unhinging experience of all. No less than the employing company, the labour union, the craft, the individual worker, was bound to the wheel of numerical 'growth', of automated, obligatory 'greed'. Shortly before he died of cancer in middle age, a dedicated British trade-union leader, Les Cannon, said: 'All I would do from now on would be based on what might seem a shocking statement—that the Trade Union Movement has lost its way, has lost its sense of mission, that the sense of brotherhood has no meaning. The powerful get what they can and they leave the weak to get what they can.'[19]

Here was a man who could still recognise incompatibles. Under the erosion of the Numbers Game they grow yearly fewer. A society compulsively dedicated, socially and individually, to *More*, cannot entertain less, even if that less is better.

A society whose dynamic derives from the race track needs losers as well as winners. When the winners mark their success with more and more costly consumption, they inevitably price out the losers. Many sub-processes of the Numbers Game combine to promote this exclusion. The force-fed multiplication of private cars leads to 'Beeching-type' dismantling of 'uneconomic' (i.e. unprofitable) public transport, escalating fares, further burdening the already most burdened—who 'fall behind' still further. The gains of the 'winners' inflate land values and rent levels, deepening the impoverishment of the others. Credit cards draw new lines of exclusion. So does the 'Leisure Industry' with its built-in compulsions. Even children are recruited to the New Poor as high pressure 'character merchandising', keyed on children's television serials, creates both 'under-privileged' children and more under-privileged parents. As a British toy manufacturer recently explained 'today every father wants to be a £5, not a 15 pence father.' In these and other ways, the Numbers Game ensures that in our 'technological consumer society', as in the primitive Palestine of the time of Christ, we shall always have the Poor with us.

In his fascinating book, *The Great Economists*, written in 1951, the American economist, Robert Heilbroner, noted that by 1950 despite rapid economic growth and a GNP per head twice that of 1939, one quarter of Americans were still living either in poverty or on its edge. However, he concluded, by the 1970s income would have grown by another 40 per cent or so in real terms, and 'by shunting more of the gains from growth towards the lower income brackets', the solution of the economic problem (which Keynes had looked forward to in the twenties) would be achieved.

The seventies have come; economic growth has been achieved yet despite an aggregation of wealth unrivalled in history, the problem of mass poverty—of what has lately been called 'The Other America'—remains as apparently chronic and pressing as ever, and in New York alone one-quarter of the population is classed as living in poverty.*

* 1 in 8 in receipt of unemployment or welfare benefit; 1 in 4 houses 'sub-standard'.

'Trapped Within a Process'

By 1971, however, faced by the unaccountable failure of the economic mechanism to respond to the control buttons according to blueprint, many economists no longer spoke with quite the old confidence. Indeed in their off-duty moments they could be as devastatingly frank as statisticians. Mr G. D. N. Worswick, director of the National Institute of Social and Economic Research, whose precise and confident forecasts regularly weigh down the headlines and send chills through British hearts, now confessed to the British Association that economics was a 'curiously disappointing science', which not only failed to provide complete answers to many central economic questions, but did not really provide any answers at all.[20] Some years before that, no less a person than Professor Simon Kuznets, the pioneer of national accounting, had raised what he called 'the fundamental question whether economics as a separate discipline can deal with the economic behaviour of such larger units [i.e. nation-states]—a question that can perhaps be parallelled by one concerning the usefulness of economics in dealing with the household unit, also much affected by major uneconomic considerations.' 'Economics,' he added, 'deals with only a part of the economic process ... and sometimes not the most important, but only the most analysable, part.'[21] This would seem to leave quite a gap.

Admittedly, since Pigou's pioneering 'welfare economics' many attempts had been made to give the discipline a sensitivity to human distinctions of quality more appropriate to its now commanding role. In the United States the National Bureau of Economic Research was seeking means by which the 'cost' of such 'disutilities' as the wreck of the landscape arising from industrial production would be taken into the national accounts. In England Mr E. J. Mishan has proposed that every citizen should be endowed with a collection of 'Amenity Rights' —to peace and quiet, space, light and so on—which an industrialist would have to buy before he could infringe them.[22] Through the web of decisions to buy and sell Rights we should thus 'optimise the aggregate utility' or even finally arrive at that mystic peak known as 'Pareto optimality'—'from which it is impossible to deviate so as to make one group better off without making some person or other group worse off.'

Unfortunately what such heroic efforts to civilise economics mainly prove is the desperate nature of the venture, and the prophetic insight of Thomas Carlyle: 'To what shifts is poor society reduced, struggling to give some account of herself, in epochs when Cash Payment has become the sole nexus between man and man.'[23]

To what shifts indeed! Yet the fact is that they are now, it seems, all we have and we are stuck with them. We appear to be trapped in the centre of a self-perpetuating politico-economic process whose dynamic derives from the maintenance of power or the maximisation of yield. In the age of psephology, it is taken as axiomatic that the trade cycle and the public opinion cycle are deliberately geared together, and that 'soft budgets' before an election will be succeeded by 'hard budgets' after it, giving way to 'soft budgets' as a new election approaches. Two British economists* have charted the 'Election Effect' on wage levels,[24] and an American lawyer, George E. Allen, has even promulgated a Law of Politico-Market Cycles, correlating stock market indices and the electoral process, which asserts that share prices always rise in the third year of a President's Administration.[25]

All in all, it was scarcely surprising to read that some American economists were moving into the field of political science with a theory which envisioned the parties as business corporations, and drew a close parallel between 'the attempt of the parties to maximise voter support' and the 'attempt of individuals to maximise utilities in the market place'. Politics were seen as 'a model of collective choice-making which is analogous to the theory of private choice embodied in the theory of the markets.'[26]

Should any flaw in the machinery develop, the cost-benefit engineers are now always available, and a leading British practitioner, Mr C. D. Foster, has explained how, by studying people's actual behaviour as an expression of their true preferences, deducing the values they attach to various benefits

---

* They calculated the average rate of increase in men's hourly earnings in non-election years 1946-1969 as 5·2 per cent; in the years running up to elections it was 7·4 per cent.[24]

—such as elbow-room on trains—and then weighting the resultant sums to register gains or losses irrespective of income group, this branch of economic science can now 'help to improve the representative quality of our democracy, to bridge the widening gap between what people want and what politicians and administrators think they want; to help the will of the majority to prevail; to take into account the interests of substantial minorities.'[27]

So, not to worry. We are all being observed and taken account of. Nevertheless, some of us do worry. In the Minority Report of Professor Colin Buchanan, the single dissentient member of the Roskill Commission, there occurs a memorable sentence: 'As the [cost-benefit research] team progressed with ever more ingenious methods of surmounting this or that difficulty or criticism, so I became more and more anxious lest I be trapped in a process which I did not fully understand and ultimately led without choice to a conclusion which I knew in my heart of hearts I did not agree with.'

This is a very contemporary feeling. For some it will seem perfectly to epitomise the citizen caught up in the cycles of the Numbers Game. But not for all. For many, as they revolve, grow accustomed to the cage, and take it for the universe.

They become the victims of the Self-fulfilling Prophecy, which lies at the heart of the Numbers Game and is the secret of its power.

THE STERILE CIRCLE

The most important point about intelligence is that we can measure it ... There is no point in arguing the question to which there is no answer, the question is what intelligence really is.

Professor Arthur Jensen, *Harvard Educational Review*, summer, 1969.

Statistical magic, like its primitive counterpart, is a mystery to the public; and like primitive magic it can never be proved wrong ... The oracle is never wrong. A mistake merely reinforces the belief in magic.

Ely Devons, *Essays in Economics*.

Among the many models of the good society, no one has urged the squirrel wheel.

J. K. Galbraith, *The Affluent Society*.

While political life was thus being eroded by the sterile imperatives of the economic Numbers Game, social horizons—and creative possibilities—were being drastically narrowed by the institutionalisation of the range of systematised self-fulfilling prophecies that spanned human life from infancy to old age.

For, unfortunately, the reaction of Professor Colin Buchanan when he found himself being 'trapped in a process which I did not fully understand' and led 'without choice to a concluson I would know in my heart of hearts I could not agree with'[1] (which was to fall back on his own lifetime of experience and insights) was now rare. More typical was a paralysed acceptance in which the very fact that one often knew, at bottom, that an 'answer' was elaborate nonsense, somehow increased its hypnotic grip and made it irresistible. Thus, a highly experienced politician and professional statistician like Harold Wilson might be well aware, in his 'heart of hearts', of the nonsense of many much-headlined political opinion poll 'gains' and 'losses'

—and yet their febrile grip could hurry him compulsively down the road to electoral disaster in 1970.

It is indeed in today's world often the most reputedly 'numerate' and hard-headed who are the Judas goats of the statistical necromancer, leading the rest of us passively to our doom. In early June 1970, London's leading financial weekly, the *Investor's Chronicle*, published an emphatic recommendation of the shares of Coalite and Chemical Products after a careful survey which underlined their sound base and the firm's expansive potential. The very next week the paper appeared with a retraction: those who had followed their advice and bought at 13s. should now sell at 12s. 9d.

What had gone wrong? Had some error in the calculations, some appalling new fact, come to light? Not at all: the fundamental case, they said, had been re-examined and remained unchanged. However, a number of 'leading technical analysts' had pointed out that in their view the chart of the share price movements which had accompanied the recommendation was showing 'a double top formation'. And a 'double top' is a selling signal. Since all those 'influential chartists' had seen the signal, the paper fairly enough pointed out that 'the shares must have an almost impossible task to make any headway, *whatever the earnings position*' (my italics).

This is a rudimentary example of this important—if largely 'invisible' phenomenon of our time—the statistical Self-fulfilling Prophecy, and further study of its many forms will show that we are entering a strange and extensive world of rational irrationality and irrational rationality and Pythagorian magic. Like the professional pornographer's, it is a largely self-sufficient world, and if its 'reality' is known, in the 'heart of hearts' to be false, it is just this, in some degree, which is the secret of its vividness and compulsive attraction.

The advertising industry has long been aware of what might perhaps be called 'the pornography of number' and has laboured mightily in laying the groundwork of statistical self-fulfilling prophecy. In his classic work on *Scientific Advertising*—and the adjective should be noted—Claude Hopkins refers to the bad old, pre-numerate days when 'manufacturers of shaving soap merely advertised vague claims like "Abundant Lather", "Acts Quickly" and so on.' Came the dawn of Scientific Advertising and a shaving soap manufacturer advertised 'Multiplies Itself in Lather 250 Times—the final result of testing 130 Formulas', while a safety razor firm advertised the '78-Second Shave'. Hopkins points the moral: 'That was definite. That man at once made a sensational advance in his sales.'

Since then statistical science plus social science plus the mass media and advertising industry have vastly enlarged and automated the system of turning self-validating statistical facts into self-fulfilling statistical prophecies. Since Kinsey, reports the American psychiatrist, Dr Rollo May, 'couples place great emphasis on book-keeping and time-table in their love-making. If they fall behind schedule, they become anxious ...'[2] Not unlike the nation 'falling behind' in that more elaborate form of self-validating 'fact', the 'international Growth League'. Opinion polls lend substance to the numerical nightmares of a Powell or a Joe McCarthy, each result consolidating a foundation in uncontrovertible figures for its successor to build on. 'The Top of the Pops', even if it really isn't, so proclaimed, may well become so. The more an 'inflationary spiral' is delineated in the headlines and those charts, the faster it is likely to spiral. The econometricians, indeed, are a fertile source of self-fulfilling prophecies. By 1959 eight European countries were following the example of the Munich Institute of Economic Research which in 1950 had started sending *monthly* questionnaires to businessmen on their intentions, *expectations*, and current outlook. Optimism is a wondrous breeder of optimism, and vague pessimism, if quantified, becomes almost a palpable feature of the landscape.* You can't argue with statistics; generally you can't even get *at* them; the strength of the self-fulfilling prophecy lies in its appearance of immaculate conception. It is a great strength. William H. Whyte tells of a large American company which gave a personality test—where 'scoring' is based on matching to a group norm—to an employee it intended to promote:

The report that the consultant firm mailed back to the company was freighted with warnings about the man's stability. The com-

* In his article on Economic Forecasting in the *International Encyclopedia of the Social Sciences*, Victor Zarnowitz, concedes that forecasts 'may influence the economic behaviour, in particular the variables, being predicted'.

pany was puzzled. The man had consistently done a fine job. Still ... The more the company mused, the more worried it became; at last it decided to tell the man the promotion he had expected so long was to go to someone else. Six months later, the company reports, the man had a nervous breakdown. As in all such stories, the company says that this proves how accurate the test was ...[3]

Such confident calculations, such indices of Expectancy established by calipers and correlation matrices, may not only put the finger on individuals, but may powerfully affect the shape and quality of whole societies. Perhaps the classic case is that of that pregnant little two-or three-number figure known as the Intelligence Quotient. Since, in its origins, style of operation, and extensive side-effects, it encapsulates nearly all the critical elements of the 57 varieties of statistical self-fulfilling prophecies that now thread our lives, we might do worse than take a look at its natural history.

Humanity Gets a New Dimension

It is no doubt appropriate that the most universal, intimate and provocative index in the whole field of the Numbers Game should lie right in the mainstream of the development of statistical science, flowing directly back to that dedicated measurer and statistician extraordinary, pioneer of meteorology and fingerprinting, 'inventor' of 'eugenics', Sir Francis Galton.

Stimulated by his cousin Charles Darwin's book on natural selection in the animal kingdom, Galton threw himself into the investigation of human inheritance—of the transmission of the marks of distinction in distinguished families. But how to define such distinction? Galton's motto echoed Kelvin: it was 'Count if you can.' Since intellectual qualities were impalpable, his efforts were directed—like those of his American disciple and one-time assistant, James McKeen Cattell—mainly to calibrating the physical senses—visual and auditory acuity, reaction times, ability to distinguish weights and colours, and so on.[4]

It was indeed in the course of his efforts to relate the attributes of parents and children that Galton stumbled upon that statistical 'open sesame', correlation. And correlation, applied to Galton's accumulated pedigrees and family records, strongly hinted at a common element, a golden core of distinction. He pioneered the study of identical twins, later parted, and from this concluded that 'nature prevails enormously over nurture'. He likened heredity to the strong current of a stream that sweeps along bits of sticks thrown in, although they might be temporarily held up by obstacles ('circumstances'). 'It is in the most unqualified manner,' wrote Galton, 'that I object to pretensions to natural equality. The experiences of the nursery, the school, the university and of professional careers are a chain of proofs to the contrary.'[5]

Although his methods were improvised and inevitably still primitive, the mark Galton left on the measurement of human ability went deep. He identified it with *innateness*, deriving from heredity. (He hardly even considered the effect of favourable material conditions on the continuing distinction of his families.) Statistically, he imported the idea of *hierarchy*. 'Scales of merit' cut his Normal curve of distribution of ability into fourteen slices, top to bottom. Finally, with his starting point in the physical or anthropometric, he saw the measure of ability as indicating a *limit* of capacity, as in a muscle—a metaphor he used—which can be strengthened by exercise, but which soon reaches the full extent of its powers.

All these ideas, with their immense political and social implications—of which Galton, the archetypal statistician, was probably not fully aware—were to be neatly packaged in the colourless, but conclusive, summations of the Intelligence Tests, soon to bulk large and fateful in so many lives.

Both the bulk and the universality—and the pregnant word 'intelligence'—derive from the French clinical psychologist, Alfred Binet, who at the turn of the century brought the notion of measuring human ability out of the Darwinian genealogical trees and the anthropological laboratory to locate it firmly in the school classroom, devising, as he did so, a new and rather intimidating statistic—the 'Mental Age'. In 1904 Binet had been invited by the French Minister of Public Instruction to serve on a commission appointed to advise on the problem of dealing with backward children in French schools. However, the first need was to identify them, and to this end Binet, in collaboration with Theo Simon, a young doctor at a mental institution, worked out a series of tests.*

'Intelligence' now became what it took to pass intelligence tests, which, in turn, were designed to predict successful performance of the exercises required in a French classroom. If a child attained the norm for his chronological age, that was his Mental Age too; if he could only succeed in the tests validated for a lower age group than his, then his Mental Age was below his real age. Not unnaturally, in view of their purpose, the tests and those later based on them, were predominantly *verbal*: a characteristic element of circularity had entered the picture.

But while the concept of 'intelligence' was thus, in practice, being tied down and narrowed, almost simultaneously the processes of statistical analysis and abstraction were raising it towards those metaphysical and universal heights which we salute in such a 'sign' as the Growth Rate. Galton had been trained as a doctor of medicine, Binet as a lawyer—and Charles Spearman was that rare, yet persistent, figure, the British army officer with an intellectual obsession. Spearman, correlating Galton's correlations, so to speak, brought the new statistical technique of 'factor analysis' to bear on test scores, laying bare with this statistical scalpel a common, underlying element which he called 'G'—for 'general intelligence'. It was accompanied, it was true, by a residue of varying amounts of specific abilities—labelled 'S'. But the mathematical analysis of Charles Spearman—who ended up a Professor of Psychology—left no doubt at all in his mind as to which was the master quality. If you were well endowed with 'G'—the message seemed to be— you might inherit the earth; if not, no use parading 'S' unless to identify yourself as a 'hewer of wood and drawer of water'.

Since these cosmic conclusions were arrived at by protracted statistical analyses which were a closed book to most people (although henceforth to occupy a major place in the assessment of human potential) a touch of the transcendental was added to the prosaic classroom ratios of the hard-working and ingenious Binet.

---

* Adapted for their own countries by Lewis Terman in the US, by Cyril Burt in Britain, and by Wilhelm Stern in Germany, the Binet-Simon tests became a sort of 'gold standard' of the educational world.

The new socio-statistical artefact was now all but launched upon its independent life. It merely remained for the German psychologist, Wilhelm Stern, to add the final touch of numerical magic by dividing Binet's 'Mental Age' by the 'Chronological Age' and multiplying by one hundred to give a crisp two or three figure quotient. 'That was definite,' as Claude Hopkins would have said. Henceforward in the advanced areas of the world, and indeed in an increasing number of the not-so-advanced, a man weighed so many pounds, stood so many inches high in his stocking feet, had a sound—or an unsound—digestion, a mortgage of such-and-such a size—and an IQ.

The weight and the mortgage he might do something about; but the height and the IQ he was lumbered with.

Predestination Brought Up-to-Date

If IQ statistics offered a new table of orientation for the human landscape, the effect in Britain at first was to confirm the familiar configuration of society.

As late as 1939 there were only 470,000 children in secondary schools in England, less than one-fifth of the age group. The wastage of ability was enormous. And yet from 1926 onwards a succession of commissions and committees had been charged with the modernisation of the nation's educational system. All had been advised by the new psychometricians. And what they heard, in effect, was that intelligence testing had revealed a large and apparently natural gulf between the 'G' of the children of the working-class mass and those of the middle-class minority. Not only did the powerful new mystique of IQ legitimise the middle-class near-monopoly of the grammar schools (and in effect secondary education), but it perpetuated a traditional tripartite division of schools, dedicated to the principle, as Raymond Williams has put it, that 'those who are slowest to learn shall have the shortest time in which to learn.'[6]

Following in the footsteps of the Hadow Committee of 1926, the Spens Committee of 1938 explained that 'intellectual development during childhood appears to progress as if it were governed by a single central factor, usually known as "general intelligence" ... Our psychological advisers assured us that it can be measured approximately ... [and that] with few exceptions, it is possible at a very early age to predict with some degree of accuracy the ultimate level of a child's intellectual powers, but this is true only of general intelligence, and does not hold good in respect of specific aptitudes or interests.'[7]

'G', however, as denoted in the IQ score, was clearly what counted, and, thus fortified, the Spens Committee continued the largely class-segregated three-school system, although now, not unaware of the changing temper of the times, it was careful to emphasise that 'parity of esteem' must exist between the grammar schools and the rest. By 1943, however, when yet another committee again rejected the 'multilateral' or common school on the American plan, the hollowness of this hollow phrase had become all too evident. In its first phase the much-acclaimed postwar 'educational revolution' in Britain was all too reminiscent of the arrangements promoted eighty years earlier by the Taunton Commission which—without benefit of scientific psychometry—simply designed a school system 'corresponding to the grades of society'.

In 1959—after almost half a century under the brave IQ banner—the Crowther Report on the education of British 15 to 18-year-olds concluded bleakly: 'most of them are not being educated.' Four years later, the Robbins Report on university education pointed with equal vigour to the persistence of a 'large reservoir of untapped ability'.

How did it come about that the measurement of the national intellectual resources which had brought these stark conclusions had, at an earlier stage, presided over an educational system which complacently accepted what in fact was a massive denial of educational opportunity?

The answer could well stand as a warning against the very real dangers of that unique combination of blinkered vision and supreme confidence in the ability to see that can be induced by dedication to statistical analysis. When the new massive IQ-testing industry was being set up in Europe, the middle-class monopoly of education in the full sense had hardly been broken. Inevitably, the tests were calibrated, standardised, validated against middle-class modes of thought and values. Test items are selected (by trial and error) so that the scores yielded will conform to the normal curve of distribution and show stability over time.[8] Statistical constructs thus become, in effect, the frame on which the extant society is draped, conferring on it a look of permanence much as a good tailor's dummy might on the cheap suit shaped around it.

It was one style of self-fulfilling prophecy. In the stance of natural scientists and buttressed by their mathematical mills, educational psychologists were measuring a situation which was, in fact, the product of a very particular economic and social structure, and they were presenting it as if it were a part of the natural order. Thus, tending to self-validation, the tests changed only slowly and partially. As late as 1960, the American, Frank Reissman, was reporting that the 'Big Three' factors in determining IQ scores were 'practice, motivation, and rapport'.[9] All built into the tests a bias in favour of the middle-class child. Formal verbal dexterity, mental parlour games of the sort dear to the hearts of intelligence testers, may appear alien and silly to working-class children. Dr Alice Heim, of the Psychology Laboratory, Cambridge, reports that 'tacit mutual contempt between tester and tested is all too frequent a phenomenon,'[10] and Professor P. E. Vernon has shown that an average of 11 points in IQ score can be added by a few hours coaching in comparable material.

Once again, what the numbers measure eclipses and removes from view—and the national consciousness—what they do not measure. And thus while the measurement of IQ did provide a passport to a wider world for a trickle of able working-class children—generally those fortunate in their families—in the shorter term its effect was to 'predict the past', casting an aura of Science and Enlightenment over the preservation of a narrow-based class privilege.

Technological change, forcing the pace, nevertheless finally compelled the nation to seek out its neglected resources of brain-power. Intelligence tests, increasingly corrected for cultural and verbal bias, became the servants of that mobilisation. Servants however, who can still get out of hand. There is at least as much threat as promise in that new, self-validating phrase, 'meritocracy'.

Statistical Abortion

Circumstances alter cases. In the United States, without the social hierarchy and class consciousness of the Old World. and possessed, furthermore, by the conviction that all things are possible, intelligence testing was embraced as a tool of democracy and educational expansion. But when one turns to the class permanently at the bottom of the social pyramid, to the Negroes, the irony of the British experience is repeated. As Gunnar Myrdal long ago pointed out in his seminal survey of the race problem, *An American Dilemma*, 'low intelligence' served for years to rationalise the denial of educational opportunity to the blacks. And it has mattered relatively little that actual intelligence-testing has shown an IQ range as wide among them as among the whites, and that, as Myrdal puts it, 'besides the recognised differences between individuals in any one group. the difference among averages of groups tend to pale into insignificance.' Alas, statistically-designed measurement operates by bulking, sampling, averaging, and in the racial situation of the United States, as earlier in the class situation of Britain, the attraction of the 'group' average—however meaningless— proves irresistible. The fact that in induction tests in the First and Second World Wars Negro recruits showed an overall average IQ 15 points below the average for white recruits*—when coupled with the notion of innateness evoked in most people's minds by IQ scores—has been enough to rationalise race prejudice and psychological *apartheid*.

How powerful can be the spell of this particular statistical artefact has now been demonstrated on an almost world scale by Professor Arthur Jensen, of the University of California who, in 1969, in an article in the *Harvard Educational Review* entitled 'Environment, Heredity and Intelligence', not only drew

* If one must play this silly 'average' game, it was also found in the First World War that Negroes from the Northern states of Ohio, Illinois, New York and Pennsylvania scored higher on average than whites from Mississippi, Arkansas, Kentucky and Georgia (Myrdal). According to A. M. Shuey, *The Testing of Negro Intelligence*, 1960, today 11 per cent of Negro school and college students score at or over the average white mean for equivalent ages. Since, by definition, 50 per cent of whites also score at or below this mean, the actual overlap is more substantial than the percentage cited suggests.

attention to the average difference of 15 IQ points—'one standard deviation'*—between American Negroes and American whites, but went on to suggest that the cause of this was largely genetic, a probability being foolishly ignored by the 'environmentalists' in American education who were, he seemed to imply, pushing against doors that were, in fact, closed.

The particular occasion of Professor Jensen's challenge was the alleged failure of the Headstart programme of 'compensatory' pre-school education for disadvantaged—often Negro— children, designed to repair in two brisk years of 'enrichment' the ravages of three centuries of slavery, terminated only a little over a century earlier (although the black infant mortality rate is still three times the white).

Reflecting on the disappointing 'IQ point gains' registered as a result of this 'crash course in Western infancy'—the phrase of a British observer[11]—Professor Jensen embarked upon statistical analysis of the test data of white, Mexican and Negro children in California schools, and emerged from his calculations with the suggestion that the Negro shortfall in IQ was about three-quarters attributable to heredity against a mere quarter to environmental factors—a long echo of the old 80:20 Nature-to-Nurture ratio firmly arrived at by Galton's disciple, Sir Cyril Burt, in England years before.

Another crisp, lucid, 'scientific' statistical artefact had now entered the scene. It, too, cast a spell. It can hardly surprise anyone—except possibly a statistician—that it was soon being brandished like the finger of God in Southern courts in answer to charges of failing to obey the Supreme Court's school desegregation ruling. Later, it aroused a storm of controversy around half the world, and, according to Jensen, stimulated at least 114 separate books and articles.

Yet what did these crisp figures, in fact, amount to? The answer is 'crisp figures': a mathematical event. Many geneticists, biologists, educational psychologists, have pointed out that such a neat, mechanical two-sided ratio cannot begin to reflect the immense intricacy, the rich, creative complexity of an

---

* As an indication of what this fearsome-sounding gulf amounts to in Britain 'a genuinely comprehensive school accepts children whose IQs span four standard deviations' (Prof. H. J. Butcher, article in *Educational Research*, February 1972).

interaction that is proceeding even prenatally, and whose constituents change and interweave so subtly that they can scarcely be separated.[12]

But in this statistics-walled world, the social scientist whose *modus operandi* are whole-heartedly statistical, may hold the keys. So life is forced into the mathematical matrix, and comes out in due course carrying the hallmark of authority. The heredity : environment ratios derive from one more model with 'constants' and 'variables' and 'x' quantities to be filled in, 'other things being equal'. Identical twins have been favoured for this purpose, since here the genetic factor is, of course, the constant, and, if they are raised apart, the share of 'environment' in the making of their IQ can, theoretically, be calculated.* If—another favoured source of material—the subjects are children brought up in an orphanage from early days, the 'environment' is the constant, and if the IQs of the fathers are known, and the IQs of the children taken, further sums become possible.

With the refinements opened up by statistical science, many varieties of complex mathematical inference on these lines can be pursued *ad infinitum*. But the ingenuity of such operations is only equalled by the opacity of the blindspots. For instance, Jensen and his British champion, Professor H. J. Eysenck, argue that if those who stress the role of environment in IQ were correct, then the scores of Negro children from higher-income groups should approach those of white children from the same environment. Showing that this doesn't happen, Jensen concludes that this again suggests the paramountcy of the hereditary factors.[13]

But what does this numerical-scale equivalency of environment amount to ? As ever, a few standard material measures of status, occupation, socio-economic class, available and addable. It leaves out everything else, including the pervasive, distorting emotions of the 'race problem' in modern America, the moral and spiritual legacies of slavery, the 'mental' environment of the black child born into a white man's world. In his devotion to statistical rigour, the psychologist, in short, leaves

* The 'answers' derived over the years since Galton launched such exercises have been far from perfect in their consistency. Readers will find much of the evidence revealingly raked over in the spring, summer and winter. 1969 issues of the *Harvard Educational Review*.

out what most people would call 'the psychological factors' in the environment.

C. D. Darlington, the British geneticist, has written: 'Environments, no more than genotypes, are to be arrayed on objective scales of value'.[14] In fact, we all know that an environment in which one individual might flourish might be disastrous for another. Again, Myrdal conveys something of the true complexity—of life as against mathematics—'No environmental stimulus has the same effect on different individuals since it affects different individuals after they have had different experiences in different succession,'[15] and Professor J. McVickers Hunt, of Illinois Psychological Development Laboratory, has stressed what he calls 'the problem of match'—of finding exactly the right environmental stimulus for a particular child at the right time.[16] If found, who can say that it will not transform the possibilities? Newton, Darwin, Churchill, Henry Ford were found dull at school: in these cases the right stimulus was found.

But however dubious the derivation and wide the margins of error,* once the numerical 'answer', ratio, score, has emerged from the statistical matrices and mills, and been committed to print it becomes an Undeniable Fact, of which even its begetter may stand in awe. Impressed by the 15 points—'one standard deviation'—difference between American Negro and white average IQs and by the 75:25 innate: acquired ratio, Professor Jensen went on to suggest that because of their demonstrated (by processing test-items) weakness in 'abstract reasoning power', Negroes might be better taught by 'associative' or rote methods—a fascinating echo of the prescription of the wartime Norwood Committee for the bulk of the British working class, whose education was to deal 'with concrete things rather than with ideas'.

And now, in democratic America, a decade after its discrediting in Britain, we even get a version of 'parity of esteem': this American 'concrete education', since it will work better for those IQ-indicated, will, we are assured, make for greater, not less, equality.

The outbursts of anger which greeted the proposal merely

* The standard error of measurement on most IQ tests, estimated by testing the same person more than once, is between 5 and 10 IQ points.[17]

represent, we are told, the fury of 'egalitarians' who are deprived by Truth of their cherished illusions. It is the sort of comment that tells more of those who make it than about those of whom it is made. For the Equality of Man is not called into question by the failure of all men to possess IQs around 100; the reference is to the dignity, uniqueness and value of the human individual and to the brotherhood of man.

Unfortunately, since this is not renderable statistically, society is now increasingly distorted to accommodate the sort of equality that is. In the advancing techno-society we could be *en route* to an IQ-contoured caste system as absolute as anything unreformed Hinduism has to offer.

Le Numéro, C'est l'Homme

No reasonably observant person would wish to deny the importance of heredity. Equally, no sensible person, however 'egalitarian', would deny that different children will need different approaches from their teachers. It is when hard, allegedly scientific, lines are drawn by obscure, prolonged and vastly complex statistical processes in which any actual children only figure, remote and disembodied, as arrays of scores, that disquiet grows. For this is to impose Statistical Diminution, and to do so in a creative situation of unknown possibilities.

Oddly enough, Wilhelm Stern, the German psychologist who 'invented' the simple numerical formula of the Intelligence Quotient, seems to have seen the danger clearly. He pleaded against over-emphasis on the number. He insisted that the child should be seen as a whole, rather than as an aggregate of parts. Alas, less than two generations later we have Professor Jensen telling us there is no point in worrying about what intelligence is; the important thing is that 'we can measure it'.

But if this *is* the concept's virtue it is equally its vice. For in today's world it virtually guarantees that the part *will* be taken for the whole. 'The [intelligence] tests,' writes Brian Jackson, Director of Britain's Advisory Centre for Education, 'are immensely useful to the bureaucracy. Thousands of complex situations are vastly simplified: situations that were better recognised and left as complex.'[18] We even get those remarkable human categories 'under-achievers' and 'over-achievers' in which a statistical abstraction—and it would be difficult to think of anything more abstract and even metaphysical than IQ—becomes somehow more real and valid than the actual child the figure is supposed to represent. According to the research study of Jacob W. Getzels and Philip W. Jackson, American children 'in the rather pejorative category of over-achievers' may be assumed to be signalling some degree of 'social psychological malfunctioning' and are apt to be 'sent to the counselling office to reduce their achievement to a level more or less in line with their IQ.'[19]

Pointing out, mildly enough, that the IQ score does not 'encompass the totality of the human mind and imagination', Getzels and Jackson conclude that 'although probably the best *single* measure we have, it nevertheless rarely accounts for more than one quarter of the variance in such crucial factors as school achievement and academic performance.' And in Britain the educational psychologist, Liam Hudson, instances boys on both the Science and Arts side of schools who possess 'unquestioned ability'—some being winners of open scholarships to Oxford and Cambridge—whose IQ scores are 'unaccountably low'.[20]

Indeed, the American psychologist, J. P. Guilford has advanced a possible explanation of such so-called 'discrepancies' by making his now famous distinction between 'convergent' and 'divergent' thinkers. Guilford's 'convergents' tend to 'retaining the known, learning the predetermined and conserving what is' while his 'divergents' tend to 'revising the known, exploring the undetermined and constructing what might be.' Since orthodox intelligence tests tend to require the unique 'correct' answer, 'divergents' who might think of several—or see the question some other way—are at greater risk. (And thanks to such mechanisms may find less and less scope for their creative imagination in society also.)

In theory, of course, the absolutism of IQ in the sense of the 'effortless superiority' conferred by Spearman's pure, all-conquering essence, 'G', has been overthrown long ago. American psychometricians, in particular, as if in continuation of the American Revolution, have shuffled and reshuffled the old hierarchy of mental attributes with great zeal. Using multiple factor analysis, Lewis L. Thurstone arrived at seven primary mental abilities, which, according to Professor D. S. Bloom, of the University of Chicago, would make 50 per cent of the American population 'gifted' in the sense of coming in the top decile of one or more of them.[21] More recently, J. P. Guilford, emerging from many years of statistical analysis of mental test papers, published a model of the intellect seen as a cube divided five ways, giving no fewer than 120 'cells'. In 1959, having statistically pinned down 'about fifty intellectual factors', he reported: 'We may say that there are at least fifty ways of being intelligent ... Simplicity,' he concluded, 'certainly has its appeal. But human nature is exceedingly complex and we may as well face the fact.'[22]

Although some of us had arrived at this conclusion quite some time ago without statistical aid, the effect is only to make the appeal of simplicity—especially when offered in neat single-number form—all the more insidious. After a survey of New York schools, M. J. Ravitz commented: 'We are slowly coming to appreciate that the real damage of the IQ test is its subtle influence on the mind of the teacher'; and as late as 1957 the Task Force on Juvenile Delinquency of the Presidential Commission on Law Enforcement found it necessary to complain that 'despite the well-known facts of instability and class-bias, most school systems and teachers still use intelligence tests as though they were stable measures of innate potential, independent of the environment ... [for this reason] teachers frequently under-estimate the ability of particular children, and scale down their expectations and levels of instruction accordingly.'

The observation is given point by the experiments of Robert Rosenthal and his colleagues, demonstrating what he has called 'the Pygmalion Effect'. Children made IQ gains, with no extra schooling, when their teachers were confidentially informed by the experimenters (incorrectly) that tests had indicated them as ripe for 'intellectual blooming'.[23]

Inside the Skinner Box

Unfortunately this is a statistical artefact, a self-fulfilling prophecy, which operates not only on the individual child and parent (who together might defy it), but on the shape and quality of society as a whole. For the 'usefulness to the bureaucracy' of IQ tests extends to the construction and provision of what has been called 'an educational system in miniature', the system of social predestination which the British call 'streaming' and the Americans, less mellifluously, but more realistically, call 'tracking' (since whereas a 'stream' flows, a track is something on which one is set, and which inexorably leads to a terminal station).

Summing up her study of streaming as practised in four English grammar schools, Dr Hilde Himmelweit reported in 1970: 'Allocation to stream is highly chancy ... Yet, once assigned, stream takes over ... [it] is an important signal to a boy of his overall worth; he adjusts his sights to it, and it affects his desire to do well.'[24]

If a boy or girl comes to the educational system from the lower socio-economic classes, the combination of home environment unfavourable to intellectual activity and the built-in cultural bias of IQ tests is likely to result in an IQ score which may grossly understate potential. It also results in allocation to a low 'track' or 'stream', and it is well established both in America and in Britain that this tends to mean to the worst classrooms, the poorest facilities, the least special attention, the weakest teachers—teachers, furthermore, to whom the poor IQ figures will, perhaps despite themselves, suggest absolute and depressingly final limits.[25] It is, as the Task Force on Juvenile Delinquency noted, exactly the reverse of what is required to repair social neglect and develop latent promise: it is 'ensuring eventual failure'.

When it is remembered that tracking or streaming may begin almost at the school entrance gates (in Britain possibly at the age of seven)* it can be seen that here is a mechanical circular

---

* Streaming is illegal in Norway and in Sweden below the age of 16 and is not used in Denmark. Abolished in Russia in 1936, it is being eliminated in Washington DC schools. Claims that it is necessary for efficient teaching have not been borne out by comparison of results in streamed and unstreamed schools in Britain. But it has been shown that Britain, which 'streams' from an early age, shows a far wider dispersal of school test scores than any of twelve other countries which keep children of all degrees of ability together until a later stage of the school life.[26]

process of depressing sterility, perpetuating social failure while suggesting inevitability and 'innateness'.

Yet, as we have seen, IQ is only one of the bland, omniscient Numbers which today not only beg the great questions, but, increasingly, efface them. A trend is a trend is a trend. Our compulsive genuflections before a specific Growth Rate, for instance, silently and repeatedly concedes—with more than a touch of the old Number Magic of Malthus—that the system and priorities on which its supposed 'necessity' is based are unchangeable, part of some natural order.

Such circular processes were increasingly characteristic of a society where the worlds—or mechanisms—of the media, of consumer industry and of politics more and more closely overlapped and meshed together. There was now a sense of completion, of logical conclusions, of the final gaps in the circles being closed.

In the early 1920s, in the opening moves of the Numbers Game, Alfred Harmsworth, having tapped the wells of consumer display advertising, marked the entry of the Press into big business by acquiring pulp forests and a paper-making plant in Canada for the *Daily Mail*. Now, half a century on in 1970, Britain's biggest selling daily, the *Daily Mirror*, and the vast International Publishing Corporation it had put together, was taken over by its paper-supplier partner, the Reed Group, a giant in paper, packaging, wallpapers, paints and do-it-yourself aids. The chairman of the Reed Group, Mr Don Ryder, who became chairman of the merged companies, told the Press that the new group was 'firmly, fervently and solidly in the consumer market ... every division of the International Publishing Corporation will have to be examined to see how it fits in.'

At the other end of the British Press spectrum, in November of the same year, Mr William Rees-Mogg, editor of the London *Times*, told a Cambridge conference on communications that the readership of *The Times* in the 'AB' (business and professional) groups had risen 55 per cent in the preceding four years, and *The Times* was thus now able to offer advertisers the second cheapest rate nationally and the cheapest rate in the South-east for reaching 'AB' men. 'This,' he explained, 'creates an identity of interests; if we can talk editorially to the audience we want to reach that also provides the future commercial strength of *The Times*.' He added that 'the very great change in *The Times* had been the result of editorial, and not commercial, aims: the result of a deliberate policy of providing a newspaper which conforms not to a market, but to an idea.'[27]

That ideas can still prevail is good news. Yet it had, clearly, to be an 'AB' idea. For those other 'streams', 'tracks', as they emerged from the educational machine, and were variously monitored and metered, making their market-research linkages with Destiny, the range in ideas—or should it be 'concepts?'—would, inferentially, be different. In 1968 that old master of the media Numbers Game, the *Saturday Evening Post*, made the point in a somewhat brutal fashion when it deliberately cut its 6.8 million circulation in half, discarding its 'C', 'D', and 'E' readers by giving them refunds, switching them to other papers, or simply not inviting them to re-subscribe. It was thus enabled to boast to its advertising agencies: 'the *Saturday Evening Post* now delivers to the better part of 3 million customer-families precisely in the 'A' and 'B' markets where advertisers account for the bulk of their sales.'[28]

How the rejected readers reacted is not recorded (any more than how the lowest 'tracks' at school feel about it). But as President Nixon's communications expert in the 1968 Presidential election, Harry Treleaven, said when asked about the new political marketing techniques: 'I don't think the audience mind the commercial approach. We've spent millions of dollars making them accept it, and now they do. They're conditioned.'

And like even the most accomplished, long-stay pigeons in one of Professor B. F. Skinner's famous Boxes, we are for most of the time unaware of the fact. As the pigeons see only the 'reinforcing' peas which reward their conformity—and not the Planner who arranges the peas' release—we, although very possibly already some little way 'beyond freedom and dignity', see only the 'reinforcing' numbers, the IQs, the Growth Rates, the Ratings, the opinion polls, the indices of the trends, the extrapolation of the curves.

THE FATEFUL PARTNERSHIP

The relation between ends and means, between the goal and the path towards it, is now seen to be far closer and more complex than has been sometimes supposed. As Marcuse writes: 'these aesthetic needs and goals must from the beginning be present in the reconstruction of society, and not only at the end or in the far future.'

> Anthony Arblaster & Steven Lukes, *The Good Society*.

'Life without the Barclaycard? It would be disastrous.'

> Housewife from Berkhamstead in the Barclaycard advertisement: 'Remember when credit was for people who couldn't handle money?'

... But that water is *composed* of two volumes of Hydrogen to one of Oxygen we cannot physically believe. Something is missing.

> D. H. Lawrence, *Phoenix*, 1936.

On the one hand there was 'modern technology', a productive machine of vast versatility and power; on the other, a pervasive sense of 'choiceless choice'—hardly relieved by the number of brands of cigarettes or 'personalised' permutations of car design, a depressing conviction that somehow almost everything was, in the phrase Fred Friendly chose as title for his book on CBS, *Due to Circumstances Beyond Our Control*. We were flatteringly assured we lived in a permissive—do-as-you-please—society. Yet when Lord Longford, leading his pornography inquiry, walked out of a live 'porno' show in Denmark in disgust, declaring that he had had as much as he could take, he was widely criticised, even in such conservative organs as the *Daily Telegraph* for thus responding to his—quite normal—feelings of revulsion. He ought, it seemed, to have proffered himself like a piece of human tissue on a slide for the effects to be 'objectively' measured.

In default of the support and authority once offered by

religion, a new sort of 'rational' or statistical puritanism now suppresses personal intuitions* and value-judgements, the things we know 'in our bones', and abjectly seeks and defers to the aggregated authority of number, whether scientific 'proof' or rating. Ours, in short, is not so much a permissive, as a sub-missive—or supine—or programmed—society.

Caught thus, between 'the logic of technology' and the statistical processing and programming described in this book, is there, in fact, any escape for us without bringing down our complex 'affluent society' in ruins? No use crying 'Stop the world, I want to get off!' as the economist and former Labour Minister, Mr Anthony Crosland, cheerfully warned disaffected socialists recently.

A diagnosis is not a cure, or even a treatment, but if the diagnosis is of Iatrogenic Disease, the remedy could be nearer at hand than we believe. A change to doctors less portentously obsessed with their instruments and more capable of seeing the whole man might work wonders. Admittedly we have been 'under' so many species of neurosis-breeding doctor so long that re-covery of our full self-confidence may take time. But as we learn at last to look beyond the dread 'indicators' and hypnotic gauges new perspectives will come into view, and we shall see that what we had been taking for 'ends' were, in fact, merely 'means', and this in itself will be a long step towards complete recovery and resumption of control of our own lives.

In the last year or two there have been signs that such a re-valuation may have already begun. It is not merely that the 'environment', 'the quality of life' and so on have been be-latedly discovered, at least as—somewhat aseptic—slogans, but also that, here and there, people have even dared to cock a snook at 'the logic of technology' and the semi-divine laws of economic 'science'. When in March 1971, the United States Congress refused funds for the continuance of the billion-dollar

---

* Much more correctly contemporary was the approach of Reuter's cor-respondent in Denmark who wrote of the country's pornography industry, 'the more I see of it, the more repulsive I find it ... because pornography reduces sex to exclusively physical manifestations and excludes feeling and commitment which are the basis of ordinary relationships ... ⌈and is⌉ in this sense degrading,' and yet declined to condemn its legalisation because 'Danish psychiatrists could find no evidence that it was harmful, even to children' and, like practically anything it might have what sociologists call a 'functional role'.

sst—supersonic transport—project, President Nixon warned that it could 'be taken as a reversal of America's tradition of staying in the vanguard of scientific and technological advance'. It could indeed—nevertheless, what had tipped the balance had clearly been the ordering, by human beings, of human priorities. Better housing, more educational opportunities, decent mass transport on the ground and so on, were felt to have the better claim. Asked why he had reversed his previous vote *for* the project, Senator Clinton Anderson, of New Mexico, replied succinctly: 'I read my mail.' The Los Angeles *Times* which had also backed the project, explained that it 'had become a symbol to a lot of people—a symbol of resistance to the so-called "military-industrial complex", a symbol of resistance to the technological spoliation of the environment . . .'[2]

And hardly less symbolic was the British Government's rejection, at about the same time, of the £1,100,000 Roskill Commission's recommendation of an inland site for London's third international airport—for again it was made under the pressure of a sort of middle-class ex-urbanite 'peasants' revolt' against the economists and their mystic, but hitherto sacrosanct, concept of 'misallocation of resources' (in this instance the high cost of building an airport on a reclaimed coastal mud-flat).[3] With no doubt deplorable 'subjectivity', poor Mr Justice Roskill was burned in effigy in the Buckinghamshire countryside the airport would have laid waste.

But such victories are precarious. If we are going to establish even a bridgehead for a breakout, we must drive down to the deeper roots of the trouble: we must diagnose the doctors before, rather than after, they diagnose us.

The Perfect Cop-Out

'The Social Science,' growled Thomas Carlyle, at the beginning of it all in the 1840s, 'not a "gay science" . . . no, a dreary, desolate and indeed quite abject and distressing one: what we might call, by way of eminence, the *dismal science*.'[4] The vast proliferation of the social sciences since Carlyle thus excoriated the political economists need in no way diminish our respect for the prophet's percipience. 'Tables,' he went on, 'are abstractions, and the object a most concrete one, so difficult to read the essence of. There are innumerable circumstances, and one left out may be the vital one on which all turned.'

Taking all human and social life to be their province, the social sciences brought a vast extension of the area where what was conceived to be 'the Scientific Method' was felt to be *de rigueur*, and this, as has been shown, was very likely to mean statistical proof. Some of the extensive side effects of these preoccupations have been described. Sometimes, while appearing to say something, the proof will have acquired its 'rigour' at the price of tautology. But often, in the nature of things, no 'scientifically valid' proof will be attainable. Then two things may easily happen. The negative 'there is no evidence' for some proposition may—society, like Nature, abhorring a vacuum—somehow become reversed to imply that it has been *dis*proved. 'There is no evidence that' pornography is 'harmful,' therefore it is good. Or, more frequently, varying degrees of interim inconclusiveness or confusion are followed by the pre-emption of 'More Research is Needed.'

In any case, 'facts'—even when phantoms—consume *mere* 'values'; the automated society gets the perfect, built-in cop-out. Because very likely what is really needed is not more research —there may have been more than enough of that all ready— but more human concern, more of the commitment from which the will to act springs. Jung has said : 'The doctor is effective only as he is affected.' But being affected is not consonant with the stance of aseptic—and in reality, self-conscious—detachment which in the social sciences is so often found to be inseparable from scientific proof. Thus, paradoxically, as the social sciences multiply today, the degree of compassion in society—and the insight that springs from it—diminishes. Barbara Wootton, who has the advantage of being a juvenile court magistrate as well as a social scientist, has aptly described the curious deadness which characterises vast acreages of juvenile delinquency research, even case study.* The delinquent, she writes, 'is never turned down by a girl, or made miserable by

---

* Theodore Roszak in *The Making of a Counter Culture* expresses the same thought more picturesquely : 'Most of our social acientists, one feels, regard the introduction of poetic vision into their work in much the way a pious monk would regard bringing a whore into a monastery.'

a bullying foreman ... or cut off from school friends because his father's job necessitated moving from Glasgow to Plymouth ...'[5]

In fact, of course, we would frequently know well enough what needs to be done—if we would care to concern ourselves as human beings, rather than experts. The accumulating, apparently insoluble 'problem of crime and delinquency' is certainly a case in point. In the United States, in particular, in the last half century a vast number of elaborate research projects have proffered answers of a diversity only equalled by their inconclusiveness. The concerned layman, seeking enlightenment from 'science', finds himself in a Byzantine maze with many entrances, abrupt turns and dead ends, but no exit.

Yet is the problem so hopelessly abstruse? As Dr D. J. West points out in his book, *The Young Offender*: 'The characteristic social background of persistent delinquents has been described *ad nauseam*.' To a great extent it is the result of monumental, cumulative—and continuing—social neglect in an automated society, shaped not by human need and human concern but, rather, remotely actuated by economic indices, ratings, market research calculations, financial yields, to which must now be added the correlation coefficients, delinquency prediction scores, Over-Achievement margins, Under-Achievement margins, personality inventory scores, and so on, of the ironically-named 'human sciences'.

The answers, in short, lie in the 'moral' area, which means in the relations between man and man.* They are not all placed conveniently on one side of the scientific social worker's desk, or at one end of the social scientist's statistical microscope. And for that reason they are today out of reach.

---

* In 1967, 7,700 people were murdered with guns in the United States and there were 126,000 robberies and assaults in which guns were employed. In the same year two million new guns were bought. It hardly requires protracted research to suggest that effective and tight gun control in the US could have a considerable impact on violent crime. But as has been vividly demonstrated by the 'law-abiding' obstruction of such a law, the real problem here is a political one, national or community.[6]

Again, in both Britain and the United States, the pressurised propagation of credit cards—in the interests of 'growth'—both vastly increases the opportunity for fraud and takes out the moral sting, reducing formerly human transactions to an abstract computer operation.

'Where There is No Vision ...'

The indignant disavowal of any element of the political, 'ideological', aesthetic or temperamental, the insistence on the immaculate conception of social-scientific truth, can be as impoverishing as the extrusion of the untidily human.

In a fascinating study of the 'objective' conclusions reached by twenty-four diverse, qualified British and American scientists on the vexed issue of the relative importance of heredity and environment in determining capacity and action, Dr Nicholas Pastore found that, with only two exceptions, the 'hereditarians' were politically conservative while the 'environmentalists' were liberals or radicals.* His sample was drawn from a mixed group of 200 psychologists, sociologists, statisticians, geneticists, anthropologists, biologists, and search among the lot failed to uncover any contradiction of the political correlation established.[7]

As ever, possible interpretations are many. But the exercise is far from of merely academic interest. Because unacknowledged values, smuggled in, so to speak, on such a scale, can install a debilitating ambiguity at the heart of contemporary society, a paralysing rectitude of behaviour, as in a bank where defalcations have long been suspected, but not yet discovered.

Every journalist, most writers, know how some single, vivid, yet apparently inconsequential, human detail may suddenly unlock the whole meaning of a situation—because now it is illuminated in the light of the imagination. Yet the more 'rigorous' the methodology, the more likely such precious insights are to be spurned as unquantifiable or because they fail to fit into the experimental design. In the name of science creative interaction between individual mind and environment is rejected in favour of a statistically derived ratio. The vital spark is conducted to earth.

In fact, had natural science or science proper proceeded

---

* Professors Eysenck and Jensen, the most recent entrants into this hoary controversy on the 'hereditarian' side, have asserted that they stand for equality of opportunity, irrespective of class, colour or race, but do not seem to have told us how they vote. Professor Eysenck's polemic, *Race, Intelligence and Education*, however, contains an interesting reference to 'the besetting sin of many people who wish to change the world.'

wholly by the rigorously inductive or experimental method found necessary to the respectability of so many social scientists, the signs are that we should still be living in a pretty medieval world. Thomas Kuhn, historian of science, has pointed out that Newton's Second Law of Gravitation had to wait for more than a century for direct quantitative investigation, when the main purpose of the measurement was to check anomalies. '... the route from theory or law to measurement,' he writes, 'can almost never be travelled backwards.'[8] Arthur Koestler, who has made a long and detailed study of the creative work of great scientists, quotes the remark of Max Planck, father of the quantum theory, that 'the pioneer scientist must have "a vivid intuitive imagination for new ideas, not generated by deduction,* but by *artistically* creative imagination.' He continues:

> In the popular imagination these men of science appear as sober ice-cold logicians, electronic brains mounted on dry sticks. But if one were shown an anthology of typical extracts from their letters and autobiographies ... and then asked to guess their professions, the likeliest answer would be: a bunch of poets or musicians of a rather romantically naïve kind. The themes that reverberate through their intimate writing are: the belittling of logic and deductive reasoning (except for verification after the act); horror of the one-track mind; distrust of too much consistency ... scepticism regarding all-too-conscious thinking ...[9]

The procedures and qualities which Koestler names these pioneer scientists as distrusting are precisely those most in evidence and felt to be most necessary in those branches of the social sciences which have so come to shape our lives. The creative dynamic is thus systematically and conscientiously removed from the heart of contemporary society, already commercially given up to automated processes. The feverish debility of Art for Art's Sake is heightened and complemented by the anaemia of Science for Science's Sake. The cumulative effect, the source of our present baffled despair, resembles that of a man furiously pedalling a bicycle in fact mounted on rollers.

* The word 'deduction' here and in the quotation below is being used in its common, rather than its technical sense (when it would correspond to 'induction', building from the particular to the general).

In addition, as the social sciences have ramified to occupy a larger and larger area of society and subject matter, each new, often narrower, specialism and methodology has tended to stake exclusive claims, particularly in the key area of psychology, competing, as the philosopher, Kathleen Nott, has put it, 'for the sole right to the definition of the human being'.[10] And each sect, as it develops its authority and scientific cachet, cultivates, cherishes, and finally fiercely guards its own characteristic blind-spots, until, almost linking, they form a necklace of opacity across the social cornea.

Putting Back What the Numbers Left Out

In seeking a cure for our iatrogenic numerical neuroses we could do a lot worse than go back to that formidable Victorian pioneer in the use of social statistics, Charles Booth, who was very much alive to their dangers as well as to their value. Booth wrote: it is 'in intensity of feeling, and not in statistics' that the 'power to move the world lies'—although 'by statistics must this power be guided, if it is to move the world aright.' He added that statistics, 'correct in themselves ... may mislead from lack of life and colour ... from the [absence of] the proportion of facts to each other, to us, to others, and to society at large.'[11]

It is this balance which we must restore. Like Booth himself, we must retain the 'feeling' and take care not to drain away the 'life and colour' which gives proportion to the whole. We must, in short, put back much that 'the numbers' have, in the name of science, extruded.

Freedom to unleash the imagination, to describe (as Booth did) as well as measure (as he also did), can bring a refreshment of vision from many quarters. 'Much of the best social science now, as in the past,' wrote that brilliant physicist and historian of science, the late J. D. Bernal, 'is found in novels and poems, in plays and films.'[12] Would a behaviourist psychologist really be ruined beyond redemption if he supplemented his measurement of sweat-output or Galvanic Skin Response with introspective—or even literary—examination of the nature of human fear? Or if the economists could, with a blast on the trumpet of Others Things Being Equal, lower the walls of their temple, if not sufficiently to observe real men rather than miniaturised computers, enough at least to touch hands with other social sciences studying human behaviour, they might yet pull back their Mystery from the verge of bankruptcy.*

But of all the things which 'the numbers leave out', perhaps the greatest loss has been that of true perspective, of the human continuity of history. 'Time-series', 'not strictly comparable', but ingeniously 'adjusted', are scarcely a substitute, and what Alvin Toffler has called 'Future Shock' is, for this reason, largely self-inflicted, a neurosis deriving from the insupportable feeling that almost every human social problem is being faced for the first time. In fact peculiarly American, and deriving from America's unique history, this unsettling notion that man is re-made by science and technology every other week is now, through the American instrumentation of the Numbers Game, spreading through the Western world.

Putting back the words and the history, realising, for instance, that Shakespeare might conceivably have a thing or two to say about 'heredity-and-environment'† overlooked by even the most refined statistical analysis, or that the people of, say, Mid-Victorian, or even Tudor, times were not wholly unacquainted with what Mr Toffler has recently christened 'Transcience' or that even 'dropping-out' is not necessarily a uniquely contemporary invention,‡ could do much to restore the dimension of depth to our sadly and dangerously two-dimensional existence.

* The Harvard Programme on Technology and Science is in fact attempting a restructuring of economics, using a team which in addition to economists includes philosophers, sociologists and political scientists. (Robin Marris: 'Towards a Reform of the Big Firm'—*New Society*, 25 September, 1970.)

† In her book, *Philosophy and Human Nature*, Kathleen Nott quotes the revealing remark of an American pupil with whom she read *Macbeth*: 'Gee, isn't it wonderful what Shakespeare knew about human nature before psychology was invented?' We can all be grateful for that.

‡ A Roman brick, found in Warwick Lane in the City of London, carries the 'scribbled' message: AYSTALIS BYS XIII VAGATIVE SIB COTIDUM, translated as 'Austalis for the last fortnight has been wandering off by himself every day.'

Man, the Second-Rate Computer?

It has to be admitted that any such reintegrative treatment faces formidable obstacles. It pits exhaustible man against inexhaustible machines. From the thirties onwards refined, speeded-up Hollerith machines have been pouring forth their mountains of statistics, classifications, samplings, correlations in overwhelming quantity; from the sixties, Charles Babbage's 'wild' dream of the calculating 'engine devouring its own tail', fulfilled to the letter, has seemed to be rushing the story outlined in this book towards its climacteric.

It is recorded that when the American explorer, Henry M. Stanley, was shown the new Maxim machine-gun in 1867 he remarked after firing 33 shots in half a minute from behind the gun's 'arrow-proof shield': 'It is a fine weapon, and will be invaluable for subduing the heathen.'[13] This almost exactly corresponds to the reaction of a great many Business School-indoctrinated managers and others to the advent of the computer. No invention, not even the steam locomotive, has ever been enveloped in so potent a mystique, compounding, as it seemed to, in its capacious interior all the other ruling numerical mystiques of our time. Certainly it most opportunely served that unstated law of applied statistics by which the precision and authority of the indices varies directly with the mileage and complexity of calculation between them and their base material: the computer achieved a separation of unprecedented length and speed.

Again, with the advent of the computer the characteristic Numbers Game eclipse of ends by means attained the proportions of caricature. In many concerns the acquisition of a computer became an end sufficient to itself, whatever its suitability to the task being performed. In a report to the British Chartered Institute of Secretaries, a computer specialist, James Allen, found that 'only a few of the five to six thousand computers at present [1971] in use in the United Kingdom have been manifestly successful ... a number have been acknowledged as disastrous.'[14] In a survey of computer installations in the United States, the management consultant, Walter J. Shroeder, found that as few as 48 per cent of the computers were being put to productive use. American businessmen were wasting £500m a year because of idle computer time and poor-quality work.

But nothing, it seemed, could halt the triumphant forward sweep of the 'high-speed' computers on their civilising mission. Each new territory annexed to the expanding statistical empire opened up new continents to penetration. In Britain even the traditionalist radio channels of the BBC now heralded 'a new breed of scholars ... the literary programmers' whose ministrations would enable us 'to cross the Great Divide between Science and the Humanities' and 'through a valuable dialogue between man and machine ... enhance beyond all recognition our understanding of the great works of literature.'

The 'literary scholar-programmer' from whose BBC talk these phrases come went on to explain:

> The human eye and brain form a notoriously fallible partnership, and, faced with a mass of information of the complexity of a literary text, encoded in the inadequate medium of information which is language, they fail to assimilate more than a proportion of the input data, and there is no guarantee that the input data will be adequately or even accurately processed.[15]

Poor old inadequately-processing 'human eye and brain'— although one might, at this point, perhaps bear in mind that it *is* the *human* eye and brain which do the reading and appreciating, in spite of—or possibly because of—its gross deficiencies, and it is not *absolutely* necessary that each brain should perceive the 'input data' in the same 'total' way.*

The point is obvious, but we neglect it at our peril. In the

---

* Statistical 'contents analysis' has been widely applied in research to US TV programmes. In 1963 Larsen, Gray and Fortis analysed 18 TV series, detecting in them 7 goals and 8 different 'methods of achievement'—further classified as 'socially approved' or 'disapproved'. Correlation of the whole complex produced 56 possible combinations of goals and methods, and, many columns of calculations later, the conclusion that there was a 'definite tendency' to project content in which 'socially approved goals were most frequently achieved by methods not socially approved.' However, again there is just no accounting for the way of the human eye and brain with its input data (as Shakespeare once observed.) '... Whether or not the television audience perceives it in this form' (i.e. according to the percentages revealed by contents analysis) concludes the research team, 'is a matter for further study.'

much-acclaimed 'partnership' between man and computer, man stands in growing danger of emerging not only as a very junior partner, but as a very second-rate computer. For since the computer cannot think in the bumbling human way, the human partner must conform to its modes of 'thought'. Confronted by the frequent spectacle of computer-made chaos, one computer management specialist characteristically accosts complaining managements with the challenge: 'What have you done to adjust to the computer? What changes have you made?' Indeed it is not unusual that a firm—and now even it seems a farm—be required to re-structure its whole organisation to match the needs of the computer.

Readers of these annals of statistical expansionism, will find the note familiar: the baffling ambiguity of a 'master-servant' relationship which seems to be constantly reversing itself, the stealthy encroachment of numerical symbols which become somehow more 'valid' than the people and things they represent. Hailed in some quarters as a shining new key to almost limitless enlargement of human choice, the enthronement of the computer, bringing instant 'feedback' from this flawless world, seemed to complete the closing of the circle, rendering the term 'human choice' superfluous.

This was now to be demonstrated on a global scale.

The Global Game

The advent of the high-speed computer more or less coincided with a new development in the theory of probability and statistics as far-reaching in some of its effects as Galton's discovery of correlation.

Three hundred years earlier, when statistical science was launched upon its course, Cardano, Bernouilli and others had deduced the Laws of Probability from the gaming table, from the fall of the dice. Now in America the eminent mathematician, John von Neumann, and his economist colleague, Oskar Morgenstern, sat in, so to speak, with the players at a poker table and derived the rules of Game Theory. As the laws of Probability made possible the development of today's natural science, Professor Morgenstern has expressed the hope that 'the understanding of far more complicated games of strategy may gradually produce similar consequences for the social sciences.'[16]

The difference, of course, is that whereas the dice are 'blind', the 'pieces' in game theory are human participants who, within the rules, react to changing circumstances and to each other's actions. And in the world of 'managerial science' the mystique of this sort of professionalised poker, heightened by the magic of the computer, has spread impressively. Soon the managerial aspirant who could not dilate upon Systems Analysis and Decision Theory was as unusual as the old-time Captain of Industry who had *not* started out as a newsboy.

But whatever the value of such elaboration in glamourising business and diverting businessmen, the fact remains that in more flexible social situations the choices open to the participants in 'games' are of a very 'de-humanised' and 'mathematical' character. In Game Theory, Professor Morgenstern authoritatively explains, it is assumed that 'the player can perform all necessary computations needed to determine his optimal behaviour' and each player is also assumed to be completely informed about 'the alternative pay-offs due to all moves made', and so on.

Thus though the 'players' may change the course of the game by their decisions, they are expected to reason rather like 'human' miniature computers. Exercises in strategy have a natural affinity with the computer. They have made the Numbers Game global, and raised the phenomenon which I have called Statistical Diminution to new heights.

Evidence of this has been accumulating at a fearful pace in recent years, notably in the United States, where by the mid-1960s politico-military 'gaming' was achieving almost industrial proportions. In over a hundred research centres between 15,000 and 30,000 scientists and officers were engaged in vast 'decision-making' games that ranged from STAGE (Simulation of Total Atomic Global Exchange), in which the computers were given 160,000 instructions, to TEMPER (Technological, Military and Political Evaluation Routine) which 'professes to simulate just about everything from the interactions of thirty-nine states in the Cold War to a nuclear exchange between two blocs in a hot one.'[17]

In its progression from shooting dice to playing poker, statistical science now produced such striking examples of statistical logic as the concepts of 'Over-Kill' or 'proved' the superior economy of 'automatic nuclear escalation'. 'The obedience of the soldier,' ruled the military judges in the trial of Lieutenant Calley for the murder of unarmed Vietnamese men, women and children, 'is not that of an automaton.' But it was, it seemed, now that of the servant of an automaton, concepts such as 'wasting the population resource' being all too characteristic products of statistical 'thinking' with its simple—now computer-assisted—erasure of unquantifiable human values.

It is true that, in an effort to take into the account the complexities of motivation, sociologists in their scores, psychologists of many persuasions, political scientists and so on, were brought into the games. But judging by the results all seem to have been forced into the mould of computer logic. In a 'counter-insurgency' game, AGILE-COIN, simulating situations in guerrilla-torn South East Asian villages, it seems that a numerical scoring scale was devised to quantify the 'loyalty' levels of the villagers. Yet the 'scores' seem to have been arrived at by a series of calculations—no doubt at high speed—in which the villagers compare the weight and wealth of the rival bidders, balance the prospect of being beaten up by the guerrillas against being rewarded by the regulars, and so on, *ad infinitum*. Neil Sheehan, formerly a *New York Times* reporter in Vietnam, has told how in 1965 the US Embassy in Saigon distributed to correspondents a Rand Corporation study which, inspired by computerised games, authoritatively concluded 'that the peasants blamed the Viet-Cong when their hamlets were blasted and their relatives killed [by American bombardment]; in effect, that shrapnel, white phosphorous and napalm were good political medicine.'[18]

The elaborate trap—perhaps the most elaborate in history—set by this sort of statistical 'rationality' has now at least been uniquely demonstrated. The war in Vietnam, the longest and most humiliating America has ever known, was also 'the fullest gamed, fullest analysed and most intensively "planned" in history.'[19]

But the matter, of course, goes much deeper than American military success or otherwise. Between the small number of affluent nations and the rest, between the former Imperial nations and their late colonial dependencies, whether in Latin America, the Middle East, or the Pacific, a deep and perilous gulf is fixed. To bridge it, as it must be bridged, requires on the side of the 'efficient' West the ability to see the world as it appears through different, but no less human, eyes. It requires the inner eye of imagination to appreciate the claim to human dignity, the suppressed fury and confusion of ancient peoples humiliated by history. It is precisely this approach which is barred by the clever-clever 'zero-sum-game' thinking of the poker-table writ global, fixed in its sterile frame of gladiatoral accountancy, your-loss-my-gain for ever and ever, amen.

The High-Speed Destruction of the Future

One might think that such experience would give pause, but this is to dangerously under-estimate the potency of the mystique of the computer in the hands of its servants. Having ravaged the present, it was turned against the even less defensible future, and Mr Herman Kahn, at the behest of the US Air Force Systems Command, computer-generated an 'array of possible futures' in the declared belief that 'the computer is free from the inhibitions and prejudices inseparable from human thinking.'[17]

It is no small thing to be deprived of the future, even when offered 'an array' of them in lieu. For in the lives of societies and the individuals who constitute them, the future has a function not unlike that of the unconscious mind, unknown, yet a presence, rich in surprises, a source of energy and creative vision. To substitute mechanical computation ('free from inhibitions' etc.) is like picking up one's 'life-style'—a computer-stick concept—from the ready-to-wear rack, as opposed to forming one's life, as a person, by living it. It is ominous, perhaps, that many appear to be no longer aware of the difference.

Not only is the capacity of computerised future-delineation for multiplying apathy and 'future shock' great, but, in the end, behind all the 'high-speed' statistical abracadabra, the probability is that, as with the weathermen, the most notable ability of such thinkers of the unthinkable will turn out to be to tell us it is wet when it is raining. Thus, the computer-powered impresario of the Great Future Show, Mr Herman Kahn, has offered on British television to bet five-to-one that Japan's annual growth rate of 10 per cent will be sustained into the future.[21] But what one would really like to know is what the computers would have projected for Japan fifty years earlier, when it was a small, largely agricultural island with only 16 per cent of its land suited to agriculture and overwhelming deficiencies in coal, oil, iron ore and almost all metals, together with one of the world's highest birth rates.

Perhaps it is fortunate for the Japanese that they could still consult the astrologers' deductions from the conjunctions of the stars as well as the charts of cost-effectiveness. For even in the most essentially quantifiable of the social sciences, intuition may in the last resort hold the key. Professor Paul Samuelson tells how Dr Robert Adams, of Standard Oil (NJ) compared different methods of economic forecasting and 'found that "being Sumner Schlichter" was about the best. But how did Schlichter do it? I could never make this out. Neither, I believe, could he.'[21]

A future that is going to be worth living is not going to be assembled from 'bits' of the present, however ingeniously shuffled and reconstructed. In his book on business forecasting Colin Hill, a former manager of Domestic Sales Analysis at the Ford Motor Company (UK) writes: 'I would suggest that if anybody had conducted a consumer survey in 1948 ... it is most unlikely that he would have sat down and designed a Volkswagen. Equally, I think it might be true that if a survey were undertaken now (in 1957) it would probably show that a Volkswagen was just what the public wanted. In both cases, the survey would be wrong.'*

Something called 'man', as well as something called 'technology', has been known to impose himself upon the future in very strange and unpredictable ways—sometimes animated by unquantifiable entities known as 'ideas' which, within the human skull, can generate extraordinary energies. Even in

* A notable case of an elaborate survey of the present failing to predict the future is that of the costly and elaborate potential user survey preceding the designing of the ill-fated Ford Edsel—a notable contrast to the first Henry Ford's approach when the public was briskly informed they could have any colour they wanted provided it was black.

these days of massive organisation research and development, a systematic survey* has suggested that about half all important environment-changing inventions—things of the calibre of the jet-engine, the pneumatic tyre, the hovercraft, the safety razor, the ballpoint pen—originated with obsessed 'little men' who drove their eccentric notions to realisation against the opposition of experts who *knew* they were impossible and of an inert public which did not see their significance. Such men, whose 'input' is no doubt highly defective, compelled 'the future' without waiting for the computer print-out, which might—or might not—have 'foreseen' them, although if given sufficient rein, it would probably have cancelled them out as 'illogical' or inconsequential.

And more rein is taken, if not given, all the time. Recently, in what looked like the apotheosis of the Numbers Game, skilfully combining every dubious aspect, the Systems Dynamics Laboratory of MIT set its computers to forecasting not merely the development of one country, but of the whole world in relation to 'the quality of life'—statistically defined and computer-assessed. 'Indicators' for five key environmental factors were linked by 45 equations in a 'world model', and the effort to extrapolate to the year 2100 via a series of computer-drawn curves proceeded.

Predictably, this exercise in 'world dynamics', statistically determined, led to an acute attack of neo-Malthusianism in which even population control appeared as self-defeating (by increasing expectations and standards of living 'growth'). A number of percentage cuts in capital investment, food production, the birth-rate and so on were proposed in order to attain the optimum (computer and index defined) 'Quality of Life' 130 years hence.

'More than 10,000 copies of the computer's findings,' it was reported, 'have been sent to high-level decision-makers throughout the world.'[22]

At the outset of this history of the Numbers Game, we saw

---

* See, for instance, *The Sources of Invention* by John Jewkes, David Sawers and Richard Stillerman, London, 2nd edition, 1969. From this study Professor Jewkes concludes: 'It might be better to recognise that all knowledge of the past, even if it could be accumulated, provides no secure base for seeing the future.'

how the statistical determinism of Malthus cast a long shadow over a century and resulted in sustained inhumanity inflicted with unshakeable self-righteousness. Could it be that the computers are regenerating that crippling nightmare, although this time the certainty and the self-righteousness (already audible) will be science- rather than religion-based, and thus—again—placed beyond effective challenge?

Counter-Attack

What we need, and need urgently, is a new sort of Declaration of Independence that would ring around the world, a declaration of independence not from territorial empire, but from techno-statistical 'logic' or imperialism. A sub-committee of the United States Senate Committee on Government Operations recently provided an excellent sketch towards Article One when it announced that 'in the making of decisions on matters of public interest, we do not propose to delegate the task to a dictator, no matter how benevolent, or to an expert, no matter how objective, or to a computer, no matter who programmes it.'*

Such attitudes today quickly attract the dismissive adjective 'Neo-Luddite'. But it is not Luddism to wish to install adequate guards on powerful machinery, or to plan to employ it in such a way that its capacity to maim its users is controlled. And of that power to maim there can now be little doubt: for, as this book has suggested, we have been brought to the point where we are in grave danger of forgetting that, as the Brazilian social anthropologist, Gilberto Freyre, has put it, 'what is human can only be explained humanely—one has to leave room for doubt, even mystery.' The objection, in short, is not to availing ourselves of the high potential of technology and statistical science, but to their totalitarian extensions which leave no room for human ambiguity, for living space. Let us have the assistance of cost-benefit analysis in complex choices by all means; but let

* The sub-committee was reporting (1967 interim report) on President Johnson's decision to employ cost-benefit accounting to locate 'national goals' and determine their relative priority 'with precision and on a continuing basis'.

us fight to the death any attempt to have the cost-benefit analysts *complete* the balance sheet. The urgent need is to regain our awareness of the limits of the Numbers Game—and mark them boldly.

At present, indeed, by excessive and undiscriminating use we are blunting valuable instruments almost beyond repair. No social scalpel can cut deeper or cleaner than a clear, proven statistic, used opportunely. Starkly recited in the House of Commons, Thomas Clarkson's cold, meticulous statistics of Negro deaths on the Middle Passage cut through clouds of sanctimonious verbiage and annihilated the powerful interests backing the Slave Trade as years of moralising argument had been unable to do. Edwin Chadwick's demonstration that typhus cost Britain as many lives in a year as the Battle of Waterloo, Charles Booth's scrupulously plotted Poverty Line offered the sort of hard-edged argument it was impossible to dismiss. Here, truly, numbers lent strength to words. But in the unending statistical stutter of our times the numbers become self-sufficient, lose impact, cease to 'connect'.

And the stutter gets worse all the time, multiplied from so many quarters. According to an American study of women's magazines, made some 98 years after Ivory Soap was discovered to be '99.44 per cent pure', the employment of percentages in their advertising pages rose by 75 per cent between 1968 and 1970.[23] Thus, *Score* made hair 12 per cent 'plumper', *Crisco* oil spattered 35 per cent less than other oils, and 45 per cent of those tested preferred Maxwell House coffee. In London a lecturer in religious education wrote to *The Times* to suggest that it was 'only a matter of time before a mathematically-minded bishop informs us that of the disciples chosen by Jesus no fewer than 8.3 per cent betrayed him.' The contemporary Tower of Babel, in short, is not a confusion of languages, but of numbers. Milton's account seems prescient:

> Forth a hideous gabble rises loud
> Among the builders; each to the other calls
> Not understood ... Thus was the building left
> Ridiculous and the work Confusion named.

If the Confusion is to be abated, a self-denying ordinance among statisticians would be a useful first step. What a relief to the spirit if we could implement the suggestion of the British economist, Professor Charles F. Carter, that 'the general principle should be to reduce all statistical information about the economy to quarterly publication (at most) unless a clear case to the contrary could be made.'* Another authority, M. J. Moroney, a senior statistician with the great international concern, Unilever Ltd, describes the proliferation of Index Numbers as 'a widespread disease of modern life', and goes on to question 'whether we would be any worse off if the whole bag of tricks were scrapped. So many are out of date, so out of touch with reality, so completely devoid of practical value ... that their regular calculation must be regarded as a widespread compulsion neurosis.'[24]

A similar critical spirit applied to the other social sciences would surely abate the authoritative headlining of so much not-fully-proven 'interim Truth'—for to blame the Press for leaving out the monumental defence-works of qualification and scholarly footnote or the public for lacking sophistication in statistical science is about as reasonable as for a doctor to complain that his patients are naïvely alarmed by his repeated recitation of their diastolic/systolic blood-pressure figures.

But looking at the wider scene, it is obvious that if we are ever to be able to use statistics to aid, rather than compel, our choices, we shall need resolute public provision. We already have consumer protection agencies for goods; now we need, even more, a consumer council for statistics, an Ombudsman of the averages and ratings, a Ralph Nader of the bland quotients and the hectoring indicators. Besides alerting the public when specific statistics conceal something very different from what they may appear, so lucidly, to be saying, such an official would insist on the printing in prominent position of standard or probable margins of error. This latter provision alone would effectively debunk many portentous statistical verdicts. Then, turning his attention to the growing encroachments of Futurology—the new focal point of our iatrogenic neuroses—he

* In *Wealth*, London, 1968. Professor Carter has made a special study of the accuracy and applications of economic statistics and was joint editor of the *Economic Journal*, 1961-1970.

might assist the recovery of the human memory (at present 'wiped' almost daily like magnetic tape) by issuing at intervals in periodical parts (with binder) straight comparisons of what was forecast and what in fact happened. Again, this should prove extremely liberating. While the prognostications of the economists would naturally provide much of the more dramatic material, the monitoring exercise would range over the whole gamut of statistical divination. No field of forecasting, for instance, is more basic and vital, and none has received more continuous and highly sophisticated attention, than that of population movements. But although grave and authoritative assertions appear in the newspapers almost weekly, the record of accuracy is dismal. In 1938 a US Presidential Commission, packed with eminent experts, doubted whether the United States' population would ever reach 140m. It was, in fact, over ten millions more only twelve years later, and is now, of course, around 200m. In 1943, after much calculation, Princeton demographers forecast a French population of only 36·9m for 1970; in fact, it is a third greater.

In the thirties and forties the busy accountants of IQ were continually issuing warnings that owing to 'the negative correlation between innate intelligence and the size of the family ... and the size of the correlation' (to quote the late Sir Cyril Burt),[25] we appeared to be facing a future of inexorably mounting stupidity. In fact, not only has the average IQ failed to decline spectacularly in obedience to these differential fertility calculations, it has even had the impertinence to tend to rise, like average height, in Britain, the US and other advanced countries.

Of course, a factor or two was left out; it generally is—as Carlyle pointed out. For at the bottom of it all is human behaviour, and human behaviour, unlike that of Professor Skinner's pigeons, remains in the long run triumphantly unpredictable.

When as a result of such restorative treatments a degree of freedom from our current abject state of statistical dependence has been obtained, it may become possible to do something about the meretricious pseudo-democratic excitements of political public opinion polls without too severe withdrawal symptoms. In Denmark, Japan, West Germany, Ontario and some other states opinion polls are now banned during the pre-election period. In Britain the Speaker's conference on the issue in 1967 recommended a ban during the final 72 hours before the close of polling; but neither of the main parties showed any disposition to dispense with the artificial, trivialising cliffhanger drama. In Britain it has also been proposed, with considerable point, that the adoption of a fixed electoral term as in the United States or West Germany, removing the Prime Minister's power to choose the strategic moment for an election (as long as he retained a majority in Parliament), would also do much to reduce the increasingly conspicuous and mechanical 'Numbers Game' element that is visibly corroding British political life.

Let us not underestimate the blanketing power of the media machine or the mind-numbing power of the statistical barrage. In today's society children need to be specially trained to resist statistical bamboozlement, inoculated against numerical neuroses as well as against smallpox and diphtheria. Infants might have fun making those authoritative graph-lines in the newspapers soar or droop by adjusting their scale (while maintaining their 'accuracy'). It is never too early to begin. Junior schools might, as an exercise combining arithmetic and sociology, unmask the easy—yet mind-paralysing—imposture of many an 'average', prodding the statistical straw of those much-loved institutions, the 'Average Man' and his 'Average Wage', and taking as text the remark of the American economist, Dr Walter Heller: 'If a man stands with his left foot on a hot stove and his right foot in a refrigerator, the statistician would say that, on the average, he's comfortable.'*

As they progress up the school system children will then learn that no lie is smoother, or more capable of staring one straight in the eye, than the statistical lie, and as an indispensable part of education for life in a 'consumer democracy' they will cultivate the ability to unmask each variety. Confronted, for instance, by that classic of beer propaganda: 'Half the

---

* Children so prepared for adult life would know how to treat the justification recently advanced by the British Post Office for a proposed diminution of the London Telephone Directory: It ran: 'a market survey indicated that the *average* subscriber in Inner London was supplied with ten times as much information as he needed.' Let us hope that the fire and hospital services do not apply the same criterion.

teetotallers in a British regiment in India died in their first year out,' they would automatically inquire the size of the teetotaller sample: there were two. They would be fully prepared for that fluent prevaricator, the 'spurious correlation', wary of extrapolation 'trips', and alert too for the wonders the distribution of emphasis can perform: they would know that 'fully ten per cent' does not differ markedly from 'only ten per cent'; that one woman in every eight reaching the age of 35 will die from cancer can equally be read that seven in every eight will not; and so on. At a later stage, it could also be immensely educational in the fullest sense to explore the human truth behind some set of statistics: it would be found, for instance, that the 'soaring crimes of violence', in Britain at least, are not necessarily, or even largely, robberies or gunpoint hold-ups, but may also be family brawls on Saturday nights.

As for political education, every election campaign should be monitored with such a manual as Darrell Huff's *How to Lie with Statistics* or W. J. Reichmann's *Use and Abuse of Statistics* in hand.

Choosing to Choose

But while such mental hygiene may enable us to distinguish from time to time the bars of our squirrel-cage from the general euphoric blur, it is unlikely to obtain our release from it. What we need, equally, as adults, is a re-direction of our scepticism, which, since T. H. Huxley invented 'agnosticism' in the wake of Darwin, has been almost exclusively directed against religion and religion-based morality. There are increasing signs, particularly in America, that, somewhat selectively, such a re-direction has already begun. At least, one hundred years after Edmund Burke announced with foreboding the onset of the age of 'sophisters, economists and calculators', the cassocks of the economists are looking decidedly frayed around the edges.

Perhaps we may yet move on from there to realise that 'Technology' is determined, as well as determining; that it is, in the words of Sir Solly Zuckerman, 'what we ourselves make of scientific knowledge ... and, unlike scientific knowledge ... which we can never reject, we can use new technology or reject it as we choose ...'[26]

We *can* choose to choose, although before we can do so we shall need to make some reorienting structural changes. If the remedy for pseudo-science is real science, with the imagination and poetry put back, the remedy for pseudo-democracy is real democracy. This will demand some re-thinking of fundamentals, and some would say, since it is 'against the trends,' that it is a hopeless aspiration. But the draining of the brain-washing bath of pseudo-science in which we now wallow will help considerably.

To seek to put back into democracy what the 'numbers' have extruded may seem to many a contradiction in terms, since democracy has come to be defined almost wholly by number,* majority, minority. But this in itself merely testifies to the unique ability of the fully deployed Numbers Game to eclipse ends by means. To free ourselves from it, and reverse this process, will demand a sustained and formidable act of will and it will exact a price in our immediate comfort. But we can still choose to choose if we will.

---

* This view of democracy in which it appears as a triumph of computation, now apotheosised by Science, is, of course, strongest in the United States. Thus, Dr George Lundberg in *Can Science Save Us?* in 1961 and Mr Alvin Toffler in *Future Shock* nine years later, advance with enthusiasm the notion of true democracy attained at last by individuals feeding their wishes into a giant computer, somewhat like children posting their requests to Santa Claus. Dr Lundberg looks forward to the time when 'through properly administered public opinion polls professionalised public officials can give us all the efficiency now claimed by authoritarian administration, and yet have the administration at all times subject to the dictates of a more delicate barometer of the people's will than is provided by all the technologically obsolete paraphernalia of traditional democratic processes.'

CAN WE RE-POSSESS THE FUTURE?

How is the human personality to maintain itself in the 'Brave New World'. On my posthumous agenda, this is, as I see it, by far the most important piece of business.

Arnold Toynbee, *Experiences*, 1969.

There is a new value system emerging in this country. For generations we have been mouthing the cliché, 'You can't stand in the way of progress.' Now there is a new generation that is saying, 'The hell you can't. That generation—and an increasing number of its elders—are saying 'Prove to us that it really is progress.'

Louis Lundberg, chairman of the Bank of America, speech at San Francisco, September 1970.

'In the past the man has been first, in the future the system must be first'—the lucid 1911 dictum of F. W. Taylor, the father of Scientific Management and of the Time-and-Motion study view of man as the 'efficient' complement to the machine, has a ring of Genesis Chapter One about it. It is not surprising that the poet and prophet of modern industrialism, Peter F. Drucker, should have pronounced Taylor's Scientific Management 'the only basic American idea that has had world-wide acceptance and impact.' It 'proved a way out of the impasse of the nineteenth century,' explains Professor Drucker, by resolving 'an irreconcilable conflict of principles', which could only have resulted in a dictatorship of either the 'capitalist' or the 'proletariat', with 'a third way'—'conflict over the division of the fruits of higher productivity.'[1]

Professor Drucker's 'third way' is roughly what has here been called the Numbers Game—with the suggestion that the freedom from the rival dictatorships may have been purchased at the cost of a more subtle and insidious subjugation. Does this, then, exhaust the possibilities open to man, or is there conceivably a fourth way, in which man, whether as worker or consumer, need not be time-and-motion studied or market-researched to feed a vastly capitalised, ever-accelerating production machine, lest its vital momentum be lost and unimaginable disaster ensue? Is it really necessary for us to be—as Mr Filmer Paradise, the American sales director of the Austin Morris car division of British Leyland, has put it—'built into a continuing relationship' with our automobile suppliers as 'manufacturers spread their net across the market, following and keeping customers as they "grow" in car-ownership and social status?'[2]

Might we, just possibly, be permitted to 'grow' some other way—by reversing the dictum of F. W. Taylor, and making man, and his true human needs, and not the System, whether capitalist or communist, the starting point? We already have —a hopeful sign of the times—the new applied science of 'ergonomics', which strives to design machinery to fit human physical and psychological characteristics, rather than vice-versa. Potentially far-reaching in its implications, it must, in practice, like that other slogan, 'participation', be severely limited by the structure and effective priorities and drives of the society in which it exists. All the same what we really seem to need is a science-art of *social* ergonomics.

But how to assess true human needs? In an 'advanced' society the answer, if obtainable, is inevitably heavily overlaid and likely to be elusive. And since, beyond the basic physical requirements, human *wants* do not exist autonomously but are heavily related to the man-made social environment and systems of expectations, the simple assumptions of the economists who equate welfare with market price paid risk a round trip. If it is to grow (in other than car ownership) human nature may have a need for altruism; but if society does not offer the opportunities, the thwarted instinct will decay. Richard Titmuss has vividly illustrated the point in his close 'case-history' of blood transfusion in Britain and America. In a similar way, emphasis on substantial money transactions in American medical practice has led, logically enough, to a vast increase in suits for damages by disappointed patients who are thus induced to put a construction on medical treatment and the doctor-patient relationship they cannot really bear. Titmuss concludes: 'If the bonds of community giving are broken, the result is not a state of value neutralism. The vacuum is likely to be filled by hostility and social conflict.'[3]

The interpretation of 'human needs' would then doubtless change. The old socialist saw about 'production for *use* instead of for profit'* would certainly be derided for its naïveté by the economist-socialists of our day. Yet, for all its woolliness, the phrase expressed a felt dilemma: that over large areas the so-called free market, working through the price system, was presenting a violently distorted view both of human needs *and* wants. Nor can it be assumed that vast computerised multiplication of choice—within the pre-set frame—will bring the reality of choice. All the signs are that, by undercutting the human base, the need to develop at a certain pace and rhythm, it will destroy meaningful choice in favour of a mindless, frenetic seizing on one contrived gimmicky craze or fashion after another, as they flow by on the conveyor belt.

Money is a poor cement of societies, and the society that cannot integrate, disintegrates. However, the social scientists have stepped into the breach. Both in Britain and America statisticians are evolving new indices of 'the quality of life', 'social indicators' to counterpoint the economic indicators and ensure, so to speak, that Mammon does not have all the best figures. In *Towards a Social Report*, published in 1969 by the US Department of Health, Education and Welfare, a panel of distinguished social scientists arrive at seven agreed statistical indicators of human welfare—measuring health, social mobility, physical environment, income level and prosperity, public order, the role of science and art, and popular participation (or its converse, alienation).

Labouring somewhat more modestly in the same vineyard, and issuing, at the end of 1970, a new periodical *Social Trends* to complement the well-established *Economic Trends*, the Director of Britain's Central Statistical Office, Mr Claus Moser, writes: 'There is a growing awareness that the fruits of economic progress can only be fully comprehended by looking at the total social effect. Providing the tools to do this is one of the greatest challenges facing the statisticians of the coming decade.'[4]

---

* Derived from William Morris: 'Now at last we will produce no more for profit but for *use*, for happiness, for LIFE' (in *News from Nowhere*).

Such 'warning lights' will have their value—always provided they can be seen in the general statistical glare. But the inherent shortcomings already discussed must make reliance on them dangerous. Can 'health', for instance, be adequately defined and measured in terms of the absence of morbidity or notified disease, or even life span? When one comes to 'the quality of life' what escapes the calipers may be vastly more important than what they encompass. Indeed, the phrase might almost be best defined by the fact of its resistance to measurement.

Mr Moser, one is relieved to note, doubts not only the practicability, but also the desirability, of a single overall social index to confront the economic GNP—a sort of GNHP or Gross National Happiness Product. But he does look forward to the creation of a system of social indicators which will not merely warn or predict, but 'through an associated system of statistical analysis, lead to an indication of the reasons for change, its consequences, and any necessary remedial actions.'

The Society of Scientists and Serfs

Francis Galton would certainly have applauded these aspirations, yet how ironic if calculations designed to restore the total welfare of the ordinary citizen to its place in the centre of the picture should end by making the man himself superfluous—except, of course, as a digit.

The danger of this ultimate conclusion to the Numbers Game is by no means so unreal as may appear. To the classic question, in its modern variant, 'Who will guard the scientists?' as they form our world, the answer tends to come back with the minimum of hesitation: 'The scientists'. Even Professor Galbraith, after his severe criticism of orthodox economics in *The New Industrial State*, nevertheless ends by resting his hopes on the scrutiny, growing numbers and involvement of 'the educational and scientific estate.'*

* Other characteristic examples: Mr Ralph Glasser, worried about the effects of marketing research, proposes a monitoring Advisory Committee on the Sociological Implications of Business Activities' 'with a strong contingent of highly qualified social scientists' and a specialist staff of social science research men.[5]

Mr Alvin Toffler, in his bestseller, *Future Shock*, suggests the need to 'develop a new profession of "value-impact forecasters" —men and women trained to use the most advanced behavioral science techniques to appraise the value implications of proposed technology.' Already, he tells us, by way of encouragement, Theodore J. Gordon, of the 'think-tank', the Institute for the Future, in Middleton, Connecticut, has pioneered 'Cross-Impact Matrix Analysis'.[6]

This is all most interesting, and it is no doubt naïve to think that before this breakthrough we already seemed to have had for quite some time a very large number of 'value-impact forecasters' and even 'cross-impact matrices,' although the latter certainly we seem to have been doing our best to break up as obsolete. The first are known as 'citizens' or, less rhetorically, as 'people', and the second are known as 'communities'; together they were once believed to form a system of 'value-impact forecasting' known as Democracy.

Of course, the people, having been trained in a great variety of schools, have a distressing tendency to arrive, without adequate methodology, at widely differing 'answers'. No doubt, as with the statistical indices, more work on them is needed. And the communities, in their tracings of cross-impact, tend to get their wires crossed. But if the memory-cores work idiosyncratically, they are nevertheless remarkably capacious and many-layered. And though often slow, the people's feedback is complex, and they possess a certain advantage in having originated the values they will be testing under impact.

Unless we are willing to settle for a neo-medieval society of planners and peasants, scientist-managers and layman-serfs —in which it is difficult to see how human values can *exist*, much less be forecast—the need is to pause and re-think 'Democracy', its ends and means. For at present it serves mainly as an ever more hollow rhetorical screen veiling a grotesque and crumbling structure in which nineteenth-century policing and sanitary administration has been superimposed on eighteenth-century legalism—and top-heavy twentieth-century technology and supermarketry on the top of that.

Such a process of re-thinking, which is likely to be protracted and exacting, would eschew alike the old élitism (whether of birth, IQ or role) carried over in the old Reithian BBC and the

the new computerised 'continuous audit' of Dr Gallup, and would go back instead to the basic ideal of democracy in the older America, placing the ordinary man at the centre, and seeking to enable him to fulfil his possibilities. Its central concern would be with values, rather than with quantities or prices.

As we grope towards such a revitalisation, we shall need to keep in the forefront three basic structural requirements.

1 We shall need to give priority to the cherishing of real communities, of human scale, whether of living-place, work-place, place of education or leisure. We shall need to do so for two reasons. First, because only thus can values be formed and proved in living, refracted through the collective experience and history; and, secondly, because only in such communities can we renew a view of democracy that rises above addition sums and psephological statistics, generating the energy and vision for the changes that are needed.

2 If the choices so reached and the values so formed are to have a chance to survive and develop, a large central area of the society and its communications must be kept free from un-remitting commercial pressures and from the mindless statisti-cal drives and self-fulfilling prophecies of the Numbers Game. Such a 'Free Zone' may appear a forlorn hope today. But once the pseudo-scientific blinkers of a self-validating, self-perpetuat-ing accountancy are shaken off, once we get outside our Skin-ner's Box, we will be able to see that there are possibilities, new ways of developing our resources, which are at present cut off from our field of vision.

3 'Pluralist' or not, a society—as distinct from a market—cannot continue without public institutions capable of em-bodying and sustaining the values that emerge and are accepted and held in common—and these may turn out to be both more numerous and simpler than in our present disintegrated state we are inclined to believe. The most consciously 'pluralist' of today's societies, the United States, nevertheless possesses a re-markably durable example of such a body in its Supreme Court which brought it to its current confrontation with the great moral issue of racial—and human—equality. But in today's conditions such institutions are unlikely to command the ac-ceptance they should until there has been a closing of the social

gulfs caused by flagrantly inequitable distribution of wealth and rewards.

It must, however, at once be admitted that in all these respects, and some others, we are currently—and have for some time been—moving rapidly in the contrary direction. Both in Britain and America local democracy has languished, getting the worst of both worlds, unable to compensate for lack of financial resources with the energy bred by genuine freedom of action. In the United States, which had once set such store by local and regional independence, great cities teeter on the edge of bankruptcy, enmeshed in the toils of an eighteenth-century Constitution which had never foreseen them. Yet the increasing encroachment of the Federal bureaucracy hardly seems the answer. In Britain, where in cities like Birmingham local government had pioneered 'Gas and Water Socialism' and had been a great nursery of Labour political leaders, local democracy was now priced out, dwarfed by developer million-aires on the one hand, and governessed on the other by White-hall with its hand on the purse-strings, and beset by the doctrine of *ultra vires*,* which, for instance, made it impossible for Bournemouth Town Council to ban dogs from its beaches without promoting a private Parliamentary Bill. Bureaucracy is well dug-in and, even more than in national politics, real debate is avoided.

Equally, on the wider, national scene, far from the area of truly public institutions being enlarged, the forum has been almost wholly taken over by the supermarket; the politics of choice have been displaced by the inexhaustible obscurantism and semi-automated processes of economics.

A Not Insoluble Dilemma

It would be foolish to pretend that current centralisation is merely the result of the power complex, or that the dilemma posed by the huge scale and soaring costs of much of the technological apparatus of modern life is not very real. But the boy who in our technological society may need to belong to an

---

* By this doctrine a British local authority cannot legally do anything it has not been specifically empowered to do.

educational area with a population of half a million in order to have his proper chance in school and college, may equally need the experience of being brought up in a genuine community of human scale and character.

If both needs were given equal attention it is not impossible that both might be met. They are not, however: Technology is enthroned. The point was vividly made by the Redcliffe-Maud Royal Commission on the modernisation of local government in Britain, which reported in 1969. Its dominating criterion for the whole vast, and immensely important exercise was the 'efficient' size for each main local authority service. Then, with its numerical requirement determined, and a measuring rod firmly in hand, the Commission proceeded to carve out of the highly varied, historically rich, intricate and living structure of England a series of areas which, regardless of whether they possessed organic unity or not, conformed as nearly as possible to the statistical model. Next to no account was taken of the creative potential of local community feeling, and its need to be nurtured rather than *further* steam-rolled. Mr Derek Senior, a journalist, the one dissenting member of the Commission who, like Professor Buchanan on the Roskill Commission, rebelled against the Efficiency shibboleth and issued a minority report, justly wrote of 'unnatural forced marriages ... not merely lacking in coherence but positively fissile.'[7]

Lancashire and Yorkshire among the counties were to cease to exist; so was Hampshire which, together with the naval town of Portsmouth, the small rural town of Petersfield, and the Isle of Wight was to be bundled into one more new 'unitary authority'. And if towns and districts which lost their old functions were to be permitted elected councils to voice local feeling to the large master authority (which must 'consult' them) this had the air of professionalism's now all-too-familiar 'public relations' humouring of the 'well-intentioned' amateur.

The change to a Conservative administration in 1970 brought, predictably, a change of accent from megalopolis to county as organisational basis, but, significantly, no less indifference to the claims of local character and democratic community traditions, the same bleak, statistical managerialism, with its overruling concern with 'financial catchment areas.' In the name of 'efficiency' even personal social services were to be transferred to the hands of the master enlarged county authorities with populations of perhaps a million or more. Famous old city and county boroughs like Oxford, Norwich, Stoke, York, Derby, with populations between 100,000 and 300,000, real living entities, strong in local patriotism and character and rich in history, the stuff of England, were to be humbled and reduced to the functions of housing management and refuse collection and minor chores. Drake's Plymouth, with a population of a quarter-million, was to be governed from Exeter, the county capital, fifty miles away. Herefordshire and Worcestershire, richly distinctive counties, created by geography and living and ruled by their own people for centuries, were to be lumped abruptly into a more efficient managerial figment labelled 'Malvernshire', and described by the indignant chairman of the —to be superseded—Herefordshire County Council as 'an unwieldy multiform administrative mass stretching from the edge of Birmingham to the Black Mountains bordering Wales'.[8] In addition, one hundred and fifty old and proud boroughs, like Stratford-upon-Avon, Canterbury, Windsor, Berwick-on-Tweed, were to vanish into the enlarged 'second-tier districts', their town halls becoming, in effect, archaeological monuments.

This way of thinking is highly infectious: no blinkers keep out the light more completely than statistical blinkers, worn in current styles. It could come as no surprise that the British Post Office had celebrated its translation into a 'modern' commercial corporation by issuing instructions that henceforth English village post offices should desist from postmarking letters with their village names—that rich and curious litany —and in the interest of 'productivity' should send their mail to the central unit for processing.

It is true that in 1969 the British Labour Party document, *Labour's Economic Strategy*, did insist that democracy required that the people must be given what it called 'opportunities to check the logic of industrial planning'. But logic is logic. Even for someone presumptuous enough to 'check the logic' of the experts, it is not a phrase which appears to offer over-much choice.

But is there not, conceivably, another approach? Instead of

constructing, statistically, vast, complete sterile structures, allegedly dictated by 'the logic of Modern Technology', can't we at least try to admit the human being—logical, or more probably, illogical—to the process at an earlier, still evolving stage?

We hear much, these days, of economic 'growth points.' But are there not also human and social growth points? Cannot we locate, and build around the nodal points in our societies where technical requirement, human operators and community life critically intersect? For instance, general practitioners in Britain's National Health Service certainly occupy such nodal points—for here meet the costly imperatives of scientific medicine, the services of public health and local government, the living and working conditions of the community, and the needs of human individuals: a microcosm of contemporary society, if seen as such. Here could be the makings of a democratic organism more sensitive to human welfare and accurately diagnostic of social weaknesses and constructive in directing their repair than any amount of detached, mechanical correlation. At the same time, the worried scientist could in this way achieve 'social responsibility,' not as the omniscient Scientist, burdened with the duty of pronouncement—à la C. P. Snow—but rather as another member of the community who, in common with others—'expert' and 'non-expert'—had a particular contribution to offer to the debate from his special experience and knowledge.

Unfortunately, what has happened is almost the reverse of this. A research study of a London borough published in 1970[9] showed the majority of GPs isolated even from the local social welfare agencies. Now the Government, seeking to introduce the approach of 'management science' here also, is establishing new health authorities which are divorced from local democracy's direct concern and manned by appointed professionals. A recent British Medical Association survey complained: 'The evaluation of medical work is made by the yardsticks of the factory floor—bed occupancy, turnover, specimens processed, and patients treated,' and warned of the mounting deficiencies in compassion and care bred by this approach.[10] (However, reassurance comes from the Business News section of the London *Times* which informs us that management consultants called in by a Sydney hospital have now constructed an objective Patient-Care Index.)

Other nodal areas where a truly participatory democracy could expose the pseudo-democracy of the Numbers Game are the meeting-points of school, community and home; the neighbourhood or 'urban village'; the workplace; the processes of town-planning and housing, even the apparently unpromising area of the relations of police and 'manor'.

Minor progress in these directions is sometimes made. While in America the Parent-Teacher Association has long been an established part of local life, in Britain until recently teachers have tended to resent what they saw as lay intrusion on their professional preserves. But since the Plowden Report on English primary schools in 1967 showed what a valuable contribution PTAS could make, they have been growing fast.

In the vexed and vital field of 'industrial democracy,' the European Economic Community has currently under discussion a new European company law, modelled on the German *Mitbestimmung* or Co-determination system, now twenty years old, which would give a company's workers the right to elect from their number one-third of its supervisory board. While this might not strike very deep, it does at least sustain the view that there are other considerations than maximum yield on capital invested and the claims of absentee shareholders.*

'Things Fall Apart ...'

But, assuming the possibility of renewal at the grass roots, the new growth today seems less and less likely to find at national level the institutions to nourish and support it. Increasingly approximating the methods—and costs—of Big Business, the national parties systematically monopolise power, effectively excluding from the electoral process the less well endowed in cash and organisational muscle. (In 1970, a Bill seeking to get some limit to the cost of seeking office in the United States by, in effect, restricting TV and radio expendi-

* Norway is introducing a similar law to come into force in 1973. Yugoslavia and Egypt among the socialist countries already operate systems of worker directors on company boards.

ture in Presidential campaigns to $5·1m for each major party, passed Congress, but was vetoed by President Nixon, whose party had spent more than twice that sum for TV and radio time in 1968. A further effort to set a ceiling on presidential campaign expenditure was similarly frustrated in late 1971.)

'The serving of public interest and the serving of the financial interest are not normally fully compatible,' Sir John Reith, architect of the BBC—and certainly no socialist—had noted mildly enough in 1932. But forty years on, in Britain where Reith's own legacy was being progressively dissipated, one was less and less aware of the existence of public institutions outside the orbit of the Numbers Game, where the common stock of values which gives any society meaning and coherence might be held, at length debated, in the Forum, rather than instantly packaged and priced in the Market. If, in C. P. Scott's phrase, an essential requisite of a newspaper was to possess 'a soul of its own,' these organs were now bought, sold, merged and made over with distressing frequency and without a by-your-leave to their communicants. The voice of the Church had grown faltering and thin since it had been established that Evil was merely a natural phenomenon statistically placed on the normal curve of distribution. The Law, which had once seemed to symbolise society's basic values, appeared increasingly circumscribed, and for some of the young a ritual object of confrontation; and the universities were caught up in various desperate numbers games of their own. The eternal values of Art were overlaid by the ticker-tape fluctuations of price and, even in England, sport had become one more business activity, now financially sponsored by brewers, oil firms, and manufacturers of cigarettes.*

While in most of its statistical foundations and some of its early manifestations the Numbers Game had its origins in Europe, its full social development and mechanisation has been mainly of American inspiration—a second American Revolution, taking off from the first and replacing the equality of man

---

* Even cricket, the last stronghold of the English amateur, was now financially sponsored by Benson and Hedges, the cigarette manufacturers. Said the marketing manager of the firm concerned: 'What we are doing is associating a brand of cigarette which has snob appeal with a sport that suits the image.' To consummate this union the firm was promoting 'brighter' one-day cricket, more suited to the TV cameras.[11]

by the equality of integers. But with the seventies there came a change. While more and more Americans recoiled from the Numbers Game, Britain, with an air of bold, modernising virtue, hastened to pursue and complete the American pattern. Commercial television had ushered in the society of the 'natural break' in 1955; now in 1972 (as America herself struggled towards a public television network) came commercial radio which had been successfully and successively resisted in Britain since the early twenties. Half a century after Coolidge, the new Conservative administration seemed in effect to be adopting the Coolidge slogan: 'The business of government is business.' One of its earliest acts was to abolish the State-backed national Consumer Council, a public watchdog organisation, at almost the same moment that in America the Senate was passing a Bill which would have created the first general Federal agency to represent and protect consumers.* The same bizarre reversal sequence was to be observed in the State organs of wage, price and profit guidance. And while in the United States, the Celler Committee delivered its Congressional onslaught on 'statistical gimmicks', the 'sophisticated accounting techniques ... making for over-statement as a way of life', of the runners-up of its conglomerate and financial empires, in Britain their counterparts remained recipients of the awed admiration felt proper to masters of the Numbers Game.

'Things fall apart; the centre cannot hold,' wrote W. B. Yeats in 1922. Half a century later things fell apart yet more spectacularly. For 'drop-out' anarchism or 'pluralism' could do little to repair the atomisation of a computerised technocracy. Repelled from one pole of abstraction, the young merely fled to another, exchanging the statistical for the psychedelic, the Economists' Equilibrium for the Buddhist Nirvana and a so-called Alternative Society of drug-aided 'reality', somewhat half-hearted communes, and self-indulgent, largely verbal, 'revolution'.

Between these opposite poles of fantasy, the middle ground crumbles. But if we no longer 'connect,' we can correlate. If we cannot envision, we can extrapolate. The statistical mills

---

* Later killed by the House Rules Committee. But Federal intervention in the protection of the consumer and the policing of the advertiser became increasingly determined.

grind faster than ever. The centre holds, but man remains divided.

A Confusion of Freedoms

Yet in the very totality of the human crisis there is a glimmer of hope. The collapse of all true urbanity in our noisome cities, the insidious failure, despite the high claims made, despite dutiful concentration on 'high productivity', to sustain 'full employment' (or to offer its alternative, properly paid leisure), the persisting, indecent gulf between rich and poor which undermines and mocks all attempts at true democracy, the built-in compulsive number-impelled inflation which so far transcends the neat little malfunction of the economists and is, in fact, the name of the game—all these in our day harshly illuminate the irreparable flaws in the Numbers Game and show that it is no longer worth playing.

Yet its failures, if we study them, clearly point to the missing elements. If we allow blind individualism as vividly symbolised in the private motor-car to dominate and wreck our cities, we do so for want of a more valid public ideal. If we tolerate mass unemployment, prescribed by indices of inflation, we do so for lack of a view that can embrace very real and heavy—but perhaps in large degree unmeasurable—social and moral costs. In the same way, progressive decay of such fundamentals of civilised life as public transport or the postal system, assisted by statistical astigmatism, underlines the central necessity of the concept of public service—with or without price and profit—as the keystone sustaining the national arch.

In the war years in Britain people hailed with enthusiasm the message of community and 'comprehensiveness' proclaimed by the construction of the National Health Service. In the years since then the market and its mindless mechanisms have increasingly eroded that vision. Yet as Britain has increasingly embraced 'the American Way'—from business schools to armoured security trucks, monopolies commissions to commercial radio—in America herself the new generation, reawakened to ideas, rejects the narrow imperatives of the Numbers Game and addresses to its elders the demand cited at the head of this chapter: 'Prove to us that it really is progress —progress for the human condition.'

This is a battle which is only just joined; a dialectic which has still far to go. On the one hand, Britain, the world's oldest and most successful representative democracy, is about to enter a European Community whose rationale is productive 'efficiency', the technological logic of the multi-national firm and of the master managers of the Brussels Commission.

On the other hand, the scepticism of the new generation in America reflects mounting scepticism everywhere, not only about the *bona fides* of technology, but also of science itself. Addressing an international conference of science-based industries at the Hague in September 1971 the former British Minister of Technology, Mr Wedgwood Benn, spoke of the 'world-wide anxiety' being felt by scientists and technologists at what he called the obvious fact that 'their long honeymoon with Governments and the people is over ... Science—and those who practise it in industry or education—is being forced to justify itself by the simple test of whether or not it appears to be serving human needs and aspirations.'

Increasingly the political cartels have encountered the anger of irregular protest and pressure groups of people not necessarily confining themselves to 'checking the logic' of technology and industry, but sometimes challenging it altogether. In the United States in mid-1970, remarking that 'the parties are becoming useless as instruments of the popular will,' John Gardner, a former Secretary of Health, Education and Welfare, formed 'Common Cause', a 'citizen's super lobby' which one year later had gathered 190,000 subscribing members, and has successfully rallied opposition to the Supersonic Transport project.[12] In Britain an embattled and highly miscellaneous group of ordinary citizens waged war on the proposal to lay waste a slice of Buckinghamshire for a great airport—and won.

Explained a spokesman of the Patients' Association, formed in Britain to combat helplessness in the face of hospital bureaucracy and medical mystique: 'I think the individual is beginning to talk back all round.' Could there be an answer to the impasse here, in Do-it-Yourself politics, a gradual accumulation of a sort of suburban 'Peasants' Revolt' against the automatism of the Numbers Game? Founded only in 1957 on the model of the us Consumers' Union, the British Consumers' Association has reached a membership of 600,000 and commands a real influence, while the circulation of the American Consumers' Union magazine, which took thirty years to reach the million mark, almost doubled in the five years to 1970. The impact of Ralph Nader and his young 'Raiders' at the significantly named 'Centre for Responsive Law' needs no emphasis.

But in these close-monitored, multi-media-ed days, 'talking back'—which might hold the possibility of creative change—all too readily passes over into mere feed-back, absorbed by the machine with hardly a tremor. 'Consumerism' merges so easily so into the all-embracing, wrap-around mass media system which invented and sustains it and which today is so largely shaped by advertising revenues and the needs of salesmanship. The process is circular and fully anaesthetised: everything approximates to the condition of showbiz.

Although it is in fact less than a long lifetime since Pulitzer on one side of the Atlantic and Northcliffe on the other built the foundation of this part of the interlocking structure of the Numbers Game by gearing advertising rates to publicly boasted certified sales figures—and considerably less than a lifetime back to American radio's invention of the competitive hourly head-count of the 'captive audience'—the whole thing has now been submitted to so much scientific survey and statistical exegesis that it appears an unalterable part of Nature, like the Record of the Rocks.

In no area is the supineness of our supine society so total. Writing of 'the universal trend to larger press units and greater monopolisation' of the Press in both Britain and America, Lord Francis-Williams, historian of the newspaper industry, added: 'It is likely to increase rather than diminish ... There is nothing to be gained by deploring this trend.'[13] (Although, in fact, as a journalist, he did deplore it.) And in 1962 yet another British Royal Commission on the Press, appointed to consider its parlous state, turned out to be mainly a resigned examination of 'market forces', dismissing every proposal for change as difficult, dangerous, or bothersome.

It is, indeed, a striking tribute to the grip the Numbers Game has attained that it should be so widely held that the Freedom of the Press can only be guaranteed by its being kept by its advertisers. That, certainly, is one form of freedom, but hardly that, perhaps, visualised by either America's Founding Fathers or the new generation, quoted by the Chairman of the Bank of America, who turn to the 'underground press', a phenomenon normally associated with repressive dictatorships.

The American banker, quoting the new generation, continued: 'What they say they want doesn't sound so different, you know, from what our founding fathers said they wanted. They said they wanted the freedom to be their own men, the freedom of self-realisation. We have lost sight of that bit in this century —but the young people are prodding us and saying: "Look, dad—this is what it's all about ..." '[14]

Today, caught up in the cushioned 'freedom' of the Numbers Game, this other sort of Freedom has its price, and it may not be low. For what we are calling 'inevitability' is, in fact, merely a refusal to recognise incompatibles, masked as they are in the colourlessness of ratio and rating. If we wish to have our newspapers—or television programmes—largely paid for by advertisers, we may get them cheaply, but we are not very likely to get them with 'souls of their own.' If the Press is to be Big Business, under competitive pressures, and the milch cow of all concerned from shareholders to journalists, it may not give first attention to its function as the witness of democracy. If a society makes More its central goal, it is naïve to complain of the workers' 'selfish' incessant wage demands or lack of 'responsibility'.

Nevertheless, the Terrorism of Numbers—particularly global 'indicators'—is not easily broken. In September 1971, faced by still lagging indices, President Nixon called for a return to 'the work ethic'; then, addressing both Houses of Congress, he asked a not long ago unimaginable question—'Do we see an America grown old and weary, past its prime, in its declining years?'.

In December new doubts were fed by the news that the Composite Index of Leading Economic Indicators had moved up only 0·2 per cent in the month. Like Britain in so many of the postwar years, America too was attacked by the creeping fear of 'lagging behind'.

In due course number-bred neuroses would give way to a number-bred euphoria hardly more real. And, then, as the indicators soared, the old question would surface: how much longer could it last? How far away was the point where number-spurred greed would give place to number-inflated fear?

Thus the Numbers Game binds us to the wheel of the contemporary fatalism, the new totalitarianism, bullying, agitating, filling our lives with mean and niggling fears and petty envies, paralysing our possibilities of free action and riveting the tyranny of 'as if there were no other scheme of things'.

There is, though.

NOTES AND REFERENCES

CHAPTER 1: THE HYPOCHONDRIACS

1 In the London *Sunday Telegraph*, 24 August 1969
2 Quoted *Time* magazine, 14 November 1969
3 Article W. A. P. Manser in *The Bankers' Magazine*, 1967
4 E. J. Mishan, 21 *Popular Economic Fallacies*, London, 1969
5 M. G. Kendall, article on History of Statistics in the *International Encyclopedia of the Social Sciences*
6 Richard Stone, *Mathematics in the Social Sciences*, London, 1965
7 Oskar Morgenstern, *On the Accuracy of Economic Observations*, Second Edition, Princeton, 1963. Table 30 computing margins of error in the calculation of growth rates
8 Health statistics—Professor Henry Miller, the *Listener*, 17 December 1970; Gerald Leach, chapter on Priorities, *The Biocrats*, London, 1970; crime statistics, Arthur M. Schlesinger, Jnr, *The Crisis of Confidence*, London, 1969
9 C. F. Carter and A. D. Roy, *British Economic Statistics*, Cambridge, 1954, an investigation commissioned by the National Institute for Social and Economic Research
10 R. G. D. Allen, *Statistics for Economists*, 3rd Edition, London, 1966
11 Norman Macrae, assistant editor of *The Economist*, in *Sunshades in October*, London, 1963. Chapter: 'The Quantity That Isn't'
12 Sir Paul Chambers, Presidential Inaugural Address to the Royal Statistical Society, 1964, reprinted RSS *Journal*, 1965

CHAPTER 2: FROM MALTHUS TO KINSEY

1 Erik Barnouw, *The Golden Web*, New York, 1966
2 Chadwick Hansen, *Witchcraft at Salem*, London, 1970
3 Parliamentary History (Hansard) 30 March 1753
4 Introduction, T. Malthus, J. Huxley, F. Osborn, *Three Essays on Population* (Mentor Books), 1960
5 Y. P. Seng, Historical Survey of the Development of Sampling Theories, *Journal of the Royal Statistical Society*, A 114
6 In Karl Pearson, *The Life, Letters and Labours of Francis Galton*, London, 1914–1930
7 Quoted Eric Roll, *A History of Economic Thought*, London, 1938
8 Walter Bagehot, *Works and Life*, Vol. VII. London, 1915
9 Harald Westergaard, *Contributions to a History of Statistics*, London, 1932
10 Donald Porter Geddes (ed.), *An Analysis of the Kinsey Reports*, New York, 1954
11 F. A. Beach, 'Characteristics of Masculine Sex Drive' in *Nebraska Symposium on Motivation* (University of Nebraska Press), 1956
12 Max Lerner, *America as a Civilisation*, London, 1958

The History of Statistics:

F. N. David, *Games, Gods and Gambling*, London, 1962
Helen M. Walker, *Studies in the History of Statistical Method*, Baltimore 1929
H. W. Macrosty, *Annals of the Royal Statistical Society, 1834–1934*, London
John Koren (ed.), *The History of Statistics*, New York, 1918
Harald L. Westergaard, *Contributions to the History of Statistics*, London, 1932
M. G. Kendall on History of Statistical Method in the *International Encyclopedia of the Social Sciences*
Launcelot Hogben, *Mathematics for the Million*, London, 1936
E. T. Bell, *Men of Mathematics*, London, 1937
Mabeth Moseley, *Irascible Genius* (Charles Babbage), London, 1964

On Insurance:

James Gollin, *Pay Now, Die Later*, New York, 1966
Harold E. Raynes, *A History of British Insurance*, London, 1964

On Medical Statistics:

Richard H. Shryock, *The Development of Modern Medicine*, 1948
Royston Lambert, *Sir John Simon*, London, 1963

On Francis Galton:

Karl Pearson, *The Life, Letters and Labours of Francis Galton*, London, 1914–1930
Francis Galton, *Memories of My Life*, London, 1908;
Chapter on Francis Galton in George A. Miller, *Psychology, the Science of Mental Life*, New York, 1962

On Malthus:

James Bonar, *Malthus and his Work*, 2nd Edition, London, 1924
J. M. Keynes, *Essays in Biography*, London, 1933
G. F. MacCleary, *The Malthusian Population Theory*, London, 1953
William Petersen, *Politics of Population*, London, 1964
Malthus, *First Essay*, with introduction by Kenneth Bolding, Ann Arbor, 1959; with introduction by Anthony Flew (Penguin), London, 1970;
Malthus, 2nd Edition, 1803, with introduction by Michael P. Fogarty (Everyman), London, 1958

On Charles Booth:

T. S. & M. B. Simey, *Charles Booth*, London, 1960
Albert Fried and R. M. Elman (eds) *Charles Booth's London*, New York, London, 1969
T. Raison, (ed.) *The Founding Fathers of Social Science* (chapter on Booth) London, 1969

Beatrice Webb, *My Apprenticeship*, London, 1926
T. S. Simey on Social Surveys, *British Journal of Sociology*, 1957
C. Booth, *The Labour and Life of the People of London*, London, 1889–1902

On Alfred C. Kinsey:

Donald Porter Geddes, (ed.), *An Analysis of the Kinsey Reports*, New York, 1954
Kinsey, Pomeroy & Martin, *Sexual Behaviour in the Human Male*, Philadelphia and London, 1948

CHAPTER 3: AMERICAN SYNCROMESH

1 Quoted in Wickham Steed, *The Press*, London, 1938
2 Quoted in Daniel Boorstin, *America, the National Experience*, New York, 1965; London, 1966
3 On the Cuban war and the Press, see W. A. Swanberg, *Citizen Hearst*, New York, 1961; London, 1962; George Juergens, *Joseph Pulitzer and the New York World*, Princeton, 1966; Frank L. Mott, *American Journalism*, New York, 1962. On Pulitzer in general: Don C. Seitz, *Joseph Pulitzer, His Life and Letters*, New York and London, 1926
4 Reginald Pound, *Selfridge*, London, 1960
  Reginald Pound and Geoffrey Harmsworth, *Northcliffe*, London, 1959
5 Francis Williams, *Dangerous Estate*, London, 1957
6 Hamilton Fyfe, *Press Parade*, London, 1936
7 Hugh Cudlipp, *Publish and Be Damned!* London, 1953
8 Lewis Jacob, *The Rise of the American Film*, New York, 1939
9 Leo A. Handel, *Hollywood Looks at Its Audience*, Urbana, 1950
10 Hortense Powdermaker, *Hollywood, the Dream Factory*, New York, 1950; London, 1951
11 Quoted in Roger Manvell, *Film*, London, 1946
12 In *The Brass Check, A Study of American Journalism*, Pasadena, 1919
13 Percival White, *Advertising Research*, New York, 1927
14 John W. Hobson, *The Selection of Advertising Media*, London, 1955
15 Gilbert Seldes, *Freedom of the Press*, New York, 1935
16 In Llewelyn White, *American Radio, a Report on Broadcasting in the U.S.A. for the Commission on the Freedom of the Press*, Chicago, 1947
17 Erik Barnouw, *Tower of Babel*, New York, 1966
18 Quoted in Gilbert Seldes, *The Great Audience*, New York, 1951
19 Paul Ferris, *Northcliffe*, London, 1971
20 C. E. Hooper and R. Chappell, *Radio Audience Measurement*, New York, 1944
21 Sam Sinclair Baker, *The Permissible Lie*, New York, 1968; London, 1969
22 Raymond Gram Swing, *Good Evening*, New York, 1964; London, 1965

CHAPTER 4: POP GOES BRITAIN

1 Sir John Reith, *Into the Wind*, London, 1949
2 Sir John Reith, *Broadcast over Britain*, London, 1924
3 Sir John Reith, *Into the Wind*, London, 1949

4 R. J. Silvey, in Edwards (ed.) *Readings in Market Research*, London, 1956
5 Sir William Haley, *The Responsibilities of Broadcasting* quoted in B. Paulu, *British Broadcasting in Transition*, London, 1961
6 Lionel Fielden, *The Natural Bent*, London, 1960
7 Maurice Gorham, *Sound and Fury*, London, 1948
8 Oliver Marriott, *The Property Boom*, London, 1967
9 Asa Briggs, *The Birth of Broadcasting*, London, 1961
10 Earl of Woolton, *Memoirs*, London, 1959
11 H. H. Wilson, *Pressure Group*, London, 1961
12 E. G. Wedell, *Broadcasting and Public Policy*, London, 1968
13 Asa Briggs, *The Golden Age of Wireless*, London, 1965
14 Burton Paulu, *British Broadcasting in Transition*, London, 1961
15 Interview in *Campaign*, 3 October 1969
16 Sir Hugh Greene, *The Third Floor Front*, London, 1969
17 David Ogilvy, *Confessions of an Advertising Man*, New York, London, 1964
18 W. F. Lynch, *The Image Industries*, London, 1960
19 *Financial Times*, 19 January 1962
20 BBC *Handbook*, 1966
21 Speech at conference 'Broadcasting and Society', at Edinburgh, reported in *The Times*, 24 March 1971

CHAPTER 5: CALCULATED OPINION

1 Chadwick Hansen, *Witchcraft at Salem*, London, 1970
2 *Time* magazine, 29 April 1971
3 Harold Wilson, article on opinion polls in *Britannica Book of the Year*, 1971
4 Harwood L. Childs, *An Introduction to Public Opinion*, New York and London, 1940
5 Walter Lippman, *Public Opinion*, New York, 1922
6 Richard Hodder-Williams, *Public Opinion Polls and British Politics*, London, 1970
7 In December 1965 the National Opinion Poll showed 63 per cent of Britons as being opposed to the existing law which made homosexual acts between consenting adults in private a criminal offence. Another National Opinion Poll showed that nearly two-thirds considered that abortion should be made legal where the mother is unable to cope with any more children, and over three-quarters where there was a serious risk that the child would be born deformed. The latter view was supported even by a majority of the Roman Catholics questioned
8 In D. Katz and others (ed.): *Public Opinion and Propaganda*, New York, 1954
9 Mark Abrams, Social Trends and Electoral Behaviour, *British Journal of Sociology*, Vol. XIII, 1962
10 Conrad Jamieson, article in the *Observer*, 9 February 1969
11 The murder rate was 3 per million population in 1957 and still 3 per million for 1968; if 'section 2 manslaughter' is included the figures become 3·5 and 4·2. This is for England and Wales; in Scotland the rate rose from 2·1 to 7·1 when the toll of murders was 37. Difficult at any time, interpretation of the statistics is greatly complicated by change in the law. Thus, anti-abolitionists point to an over 100 per cent increase in murders formerly classed as capital, following total abolition

in 1964—from 71 for the 4-year period 1960-64 to a total of 161 for the period 1964-68. Abolitionists, however, point out that juries were formerly reluctant to convict in capital murder—and the largest increase in convictions for murder was in a type of murder 'akin to manslaughter'. (Home Office Report, Murder, 1957-68)

12 George A. Miller, Psychology—The Science of Mental Life, 1962: 'Apparently some people like to say yes, no matter what you ask them, and other people like to say no'

13 David Butler and Donald Stokes, Political Change in Britain, London, 1969

14 Hugh Parry and Helen M. Crossley in Public Opinion and Propaganda, (ed.) D. Katz, 1954

15 Quoted in Leo A. Handel, Hollywood Looks at Its Audience, New York, 1950

16 H. Cantril, Gauging Public Opinion, Princeton, 1944

17 Report of the Commission on Obscenity and Pornography, New York, 1970

18 R. Jowell and G. Hoinville, 'Public Opinion Polls Tested', New Society, 7 August 1969; research commissioned by the Socialist Commentary

19 Speech at Walsall in August 1969. But a speech at York in early September 1968 shows that Heath, having noted the insistent message of the opinion polls, was already moving strongly in this direction

20 George Gallup, Guide to Public Opinion Polls, Princeton, 1948

21 Survey of Race Relations by Dr Mark Abrams, included in condensed form in E. J. B. Rose and associates, Colour and Citizenship, an Institute of Race Relations Report, Oxford, 1969

22 Quoted in R. Segal, America's Receding Future, London, 1948

23 Fred W. Friendly, Due to Circumstances Beyond Our Control, New York, London

CHAPTER 6: UNTOUCHED BY HUMAN HAND

1 D. M. Goodacre, The Use of a Sociometric Test as a Predictor of Combat Effectiveness, 1951, found that 'a high rate of choosing within a group was associated with a high standard of group effectiveness; reduced to the simplest interpretation it appears that if members of the group consider each other good for the operation of the group, the group is likely to be successful.' Quoted E. F. Borgatta in 'Sociometry' in the International Encyclopedia of the Economic and Social Sciences

2 Report of the Commission on Obscenity and Pornography, Part IV, New York, 1970

3 S. S. and Eleanor Glueck, Delinquents and Non-Delinquents, Harvard University Press, 1959

4 Hermann Mannheim and Leslie T. Wilkins, Prediction Methods in Relation to Borstal Training, HMSO, London, 1955

5 John P. Conrad, Crime and Its Correction, California and London, 1965

6 Time magazine, 18 January 1971

7 Howard Jones, Crime in a Changing Society, London, 1967

8 In A. T. Welford (ed.) Society—Methods of Study, London

9 Probation Research: a Preliminary Report, HMSO, London, 1966

10 Dr Steven Folkard, Probation Research, techniques and results in Frontiers of Criminology, (ed.) Hugh J. Klare and David Haxby, London, 1967

11 Hans Zeisel, *Say It With Figures*, New York, London, 1947
12 Report extracts quoted in J. W. Krutch, 'Through Happiness with Slide-rule and Calipers', *Saturday Review*, 2 November 1963

CHAPTER 7: *VACUUM-PACKED*

1 UNESCO Reports and Papers on Mass Communications—No. 43: *The Effects of Television on Children and Adolescents*
2 H. L. Himmelweit et al, *Television and the Child*, Oxford, 1958
3 J. D. Halloran et al, *Television and Delinquency*, Leicester, 1970
4 Sir Solly Zuckerman, *Beyond the Ivory Tower*, London, 1970
5 T. S. Simey, article on Social Investigation, *British Journal of Sociology*, Vol. 8, 1957
6 In Leon Arons and Mark May (ed.), *Television and Human Behavior*, New York, 1963
7 Research by Dr Henry K. Beecher of Harvard, quoted in Brian Inglis, *Drugs, Doctors and Disease*, London, 1965
8 Robert Rosenthal, *New Society* article, 7 November 1968
9 Leonard Berkowitz, in *Scientific American*, February 1964
10 In O. N. Larsen (ed.), *Violence and the Mass Media*, New York, 1968
11 Erna Christiansen in *Children and TV*, Washington, 1967
12 Barbara Wootton, *Social Science and Social Pathology*, London, 1959
13 Fredric Wertham, *A Sign for Cain*, New York, 1966, London, 1968
14 Fredric Wertham, in the *American Journal of Psychology*, October 1962
15 Report of the Commission on Obscenity and Pornography, New York, 1970
16 David Robinson, 'Trevelyan's Social History: Some Notes and a Cronology', in *Sight and Sound*, spring 1971
17 There has been a good deal of partisan controversy over the meaning, if any, of these statistics, but the gist of the matter—including the absurd prematurity—seems clear enough. See Commission's Minority Report (by Hill-Link-Keating) on the Danish psychologist Kutschinsky's research study; Letter from Copenhagen by Reuter's correspondent, J. J. Barnes in *Encounter*, September 1970; Ernest van den Haag, in *Encounter*, August 1970 and December 1967; Report by Raymond P. Gauer, printed in the *Congressional Record*, spring 1970

CHAPTER 8: *THE OPIATE OF OBJECTIVITY*

1 *Sunday Dispatch* (London) 10 July 1960
2 Quoted in O. N. Larsen (ed.) *Violence and the Mass Media*, 1968
3 Gunnar Myrdal, *Objectivity in Social Research*, London, 1970
4 B. Lander, *Towards an Understanding of Juvenile Delinquency*, New York, 1954
5 B. Fleisher, *The Economics of Delinquency*, Chicago, 1966
6 Appendix to the Report of the National Commission on Law Enforcement, 1965
7 Report of the Commission on Obscenity and Pornography, including Part IV, Separate Statements by Commission Members, New York, 1970
8 Fredric Wertham, *A Sign for Cain*, New York, 1966, London, 1968
9 Quoted in O. N. Larsen (ed.) *Violence and the Mass Media*, 1968

10 P. F. Lazarsfeld, *The Academic Mind: Social Scientists in Time of Crisis*
11 Reinhardt Bendix, *Embattled Reason—Essays in Social Knowledge*, New York, 1970
12 In D. Lerner and H. D. Lasswell, *The Policy Sciences*, Stanford, 1951
13 E. P. Thompson (ed.), *Warwick University Ltd*, London, 1970
14 Alvin Toffler, *Future Shock*, New York, London, 1970
15 Quoted in Francis Williams, *The Right to Know*, London, 1969
16 Joseph Seibert and Gordon Willis, Introduction in *Marketing Research: Readings*, London, 1970
17 Ralph Glasser, *Planned Marketing*, 1964; *The New High Priesthood*, London, 1967
18 Graham Bannock, *The Juggernauts*, London, 1971
19 David Ogilvy, *Confessions of an Advertising Man*, New York, London, 1964
20 Pierre Martineau, *Motivation in Advertising*, London, 1957
21 Ministry of Agriculture, Committee on Food Standards.
22 Sam Sinclair Baker, *The Permissible Lie*, New York London, 1969
23 Jacques Barzun, *Science, the Glorious Entertainment*, New York and London, 1964
24 Beatrice Potter, *My Apprenticeship*, Appendix—On the Nature of the Economic Science, London, 1926

CHAPTER 9: HANDS—INVISIBLE, VISIBLE AND DEAD

1 F. Fraser Darling, The Reith Lectures, 1969
2 Wassily Leontief, in *Quantity and Quality*, (ed.) D. Lerner 1961
3 Eric Roll, *A History of Economic Thought*, Revised Edition, London, 1961
4 H. Theill, *Applied Economic Forecasting*, 1966; C. E. V. Leser, *Can Economists Foretell the Future?* Leeds University Press, 1969
5 Michael Stewart, *Keynes and After*, London, 1967
6 H. O. Stekler, Forecasting with Economic Models, an Evaluation, *Econometrics*, Vol. 36, 1968
7 Quoted in E. W. Streissler, *Pitfalls in Econometric Forecasting*, London, 1960
8 Oskar Morgenstern, *On the Accuracy of Economic Observations*, Second Revised Edition, Princeton 1963
9 Ely Devons, *Essays in Economics*, London, 1961
10 C. E. V. Leser, *Can Economists Foretell the Future?* Leeds University Press, 1969
11 Sir John Figges in *Japan, Miracle '70*, London, 1970
12 J. K. Galbraith, *The New Industrial State*, Boston, London, 1967
13 State of the Union Message, February, 1970
14 David Bonavia in the London *Times*, 10 August 1970
15 John Davis, in the *Observer*, 22 January 1967; Malcolm Burne, 'Woolworths Tries to Change its Image', *Sunday Telegraph*, 18 April 1971
16 Interview, London *Times*, 18 September 1969
17 *Modern Farming and the Soil*, HMSO, London, 1971
18 Alan Day in the *Observer*, 12 May 1963
19 Peter Hall in *Labour's New Frontiers*, (ed.) Peter Hall, London, 1964
20 A. R. Prest and R. Turvey, 'Cost Benefit Analysis: Survey' in *Economic Journal*, December 1965; also G. H. Peters, *Cost-Benefit Analysis and Public Expenditure*, Eaton Papers, London, 1968
21 *The London–Birmingham Motorway*, Road Research Laboratory, 1960

22 Samuel B. Chase, *Problems in Public Expenditure Analysis*, Washington D.C., 1968
23 R. F. F. Dawson, *Cost of Road Accidents in Great Britain*, Road Research Laboratory, 1967
24 In Samuel B. Chase (ed.), *Problems in Public Expenditure Analysis*, Washington D.C., 1968
25 C. D. Foster, *The Transport Problem*, London, 1963
26 Gerald Leach, 'The Life Industry', in the *Observer Review*, 26 April 1970
27 Report of the Roskill Commission on the Third London Airport, HMSO, 1970
28 Sir Solly Zuckerman, *Beyond the Ivory Tower*, London, 1970
29 J. M. Keynes, *Essays in Persuasion*, London, 1931
30 Richard Titmuss, *The Gift Relationship*, London, 1970
31 'Inflation : Worst Crisis since 1931' in the *Observer*, 15 November 1970

CHAPTER 10: WHAT PRICE DEMOCRACY?

1 Andrew Alexander and Alan Watkins, 'The Making of the Prime Minister', *Sunday Telegraph*, June 1970
2 David Butler and Michael Pinto Duschinsky, *The British General Election of 1970*, London, 1971
3 Joe McGinniss, *The Selling of the President*, New York, 1969, London, 1970, quoting John Maddox
4 James M. Perry, *The New Politics*, London, 1968
5 D. E. Butler and R. Rose, *The British General Election of 1959*, London, 1960
6 Mark Abrams and Richard Rose, *Must Labour Lose?*, London, 1960
7 Anthony Howard and Richard West, *The Making of the Prime Minister*, London, 1965
8 A. E. Cooper, MP, in letter to *The Times*, 28 October 1965
9 In letter to *The Times*, 28 October 1965
10 Richard Hodder-Williams, *British Opinion Polls and British Politics*, London, 1971
11 On British TV May, 1970
12 Butler and Duschinsky, *The British General Election of 1970*
13 In 'Panorama', transcript, the *Listener*, 19 October 1967
14 Correspondence, in *The Times*, 20 January 1969
15 On BBC TV, 21 May 1969
16 *The Times*, 3 May 1969
17 Report of Industrial Dispute Inquiry, 6 November 1969
18 Quoted by Anthony Sampson 'The Hot Seats of Learning', the *Observer*, 18 October 1970
19 In BBC Nicholas Woolley interview, *Listener*, 28 January 1971
20 G. D. N. Worswick, presidential address to economic section of the British Association, 2 September 1971
21 In D. Lerner (ed.) *Parts and Wholes*, the Haydon Colloquium of the Scientific Method and Concept, New York, 1963
22 E. J. Mishan, *Growth: The Price We Pay*, London, 1969
23 In *Chartism*, 1840
24 Professor H. A. Turner and D. A. S. Jackson, article and chart in *The Times*, 24 March 1970
25 *Time* magazine, 15 February 1971
26 James M. Buchanan and Gordon Tulloch, *The Calculus of Consent*, Ann Arbor, 1962

27 Christopher Foster, 'Beeching and Beyond', in the *Listener*, 23 May 1963

CHAPTER 11: *THE STERILE CIRCLE*

1 Report of the Roskill Commission on the Third London Airport: Professor Buchanan's Note of Dissent
2 Rollo May, *Love and Will*, London, 1969
3 W. H. Whyte, *Organisation Man*, New York, London, 1957
4 George A. Miller, *Psychology, the Science of Mental Life*, New York, 1962, London 1964
5 Francis Galton, *Hereditary Genius*, London, 1892
6 Raymond Williams, *The Long Revolution*, London, 1961
7 Report of the Consultative Committee on Secondary Education, HMSO 1938
8 K. Richardson and D. Spears (ed.), *Race, Culture and Intelligence*, London, 1972: chapter 'IQ, the Illusion of Objectivity' by Joanna Ryan
9 Frank Reissman, 'The Culturally Deprived Child' (1962) quoted in report Task Force on Juvenile Delinquency, Appendix, President's Commission on Law Enforcement, 1967
10 Alice Heim, *Intelligence and Personality*, London, 1970
11 Ruth Adam, 'Project Headstart', *New Society*, 30 October 1969
12 *Race, Culture and Intelligence*
13 H. J. Eysenck, *Race, Intelligence and Education*, London, 1971
14 C. D. Darlington, *Genetics and Man*, London, 1966
15 Gunnar Myrdal, *An American Dilemma*, Vol. I
16 J. McVickers Hunt, 'Intelligence and Experience', in S. Wiseman (ed.), *Intelligence and Ability*, London, 1967
17 Professor A. R. Jensen, article, *Harvard Education Review*, 1969
18 Brian Jackson and Denis Marsden, *Education and the Working Class*, London, 1962
19 J. W. Getzels and P. W. Jackson, *Creativity and Intelligence*, London, New York, 1962
20 Liam Hudson (ed.) *The Ecology of Human Intelligence*, London, 1970; also *New Society*, article, 15 November 1962
21 Letter in *Harvard Educational Review*, spring, 1969
22 J. P. Guilford, 'Three Faces of the Intellect', in S. Wiseman (ed.) *Intelligence and Ability*, London, 1967
23 R. Rosenthal and L. F. Jackson, *Pygmalion in the Classroom*, New York, 1968
24 Dr Hilde Himmelweit, letter in the *Observer*, 7 February 1970
25 Brian Jackson, *Streaming, an Educational System in Miniature*, London, 1964; J. W. B. Douglas, *The Home and School*, London, 1964; recent research by Dr C. B. Hindley and Miss J. A. Munro at the Centre for Human Development, London, reported in the *Observer*: 'Workers' Children Lose Out on IQ', 4 July 1971; and work of Heber and Garber in the US in demonstrating the effects of special stimulation programmes on the IQs of black slum infants (3 January 1972)
26 Douglas Pidgeon, 'The Expanding Mind', *Sunday Times Magazine*, 2 March 1969
27 *The Times*, 14 November, 1970
28 J. L. Hultang and R. P. Nelson, *The Fourth Estate*, New York, 1971

CHAPTER 12: THE FATEFUL PARTNERSHIP

1 H. J. Barnes, Reuter's correspondent in Denmark, in *Encounter* September 1970
2 *Time* magazine, 5 April 1971
3 'Automatic priority,' the Commissioners said, 'should not be given to economic growth, but without it they could not see how to pay for an improved environment.'
4 Thomas Carlyle, *The Nigger Question*, London, 1849
5 Barbara Wootton, *Social Science and Social Pathology*, London, 1959
6 FBI figures
7 Nicholas Pastore, *The Nature–Nurture Controversy*, New York, 1949
8 Conference on Quantification in the History of Science, reported in *Isis*, June 1961
9 Arthur Koestler, *The Act of Creation*, London, 1964
10 Kathleen Nott, *Philosophy and Human Nature*, London, 1970
11 Charles Booth, *Life and Labour of the People of London*, 1889: also quoted in T. S. Simey, article on 'Social Investigation' in *British Journal of Sociology*, Vol 8, 1957
12 J. D. Bernal, *Science in History*, Vol IV: 'The Social Sciences', London, 1965 Edition.
13 *Illustrated London News*, report and picture.
14 Quoted Leith McGrandle, 'Is Your Computer Really Necessary?' *Sunday Telegraph*, London, 18 April 1971
15 R. W. Last, 'The Computer and the Critic', in the *Listener* 8 October 1970
16 Oskar Morgenstern, article on Game Theory in the *International Encyclopedia of the Social Sciences*.
17 Andrew Wilson, *The Bomb and the Computer*, London, 1968
18 Neil Sheehan, 'After Calley—American on Trial' in *Sunday Telegraph*, 4 April 1971
19 Andrew Wilson, *War Gaming*, London, 1970
20 Interview with Bernard Levin, ITV, reprinted in *The Times*, 29 December 1970
21 Paul A. Samuelson, 'Economic Forecasting and Science' in the *Michigan Quarterly Review*, October 1965
22 *The Futurist*, August 1971; article, 'Shock Findings on the Environment Crisis' by Gerald Leach, *Observer* 27 June 1971; Denis D. Meadows, *The Limits of Growth*, New York, London, 1972
23 *Time* magazine, 12 October 1970
24 M. J. Moroney, *Facts from Figures*, Revised Edition, London, 1956
25 Sir Cyril Burt, *Intelligence and Fertility*, London, 1946
26 Sir Solly Zuckerman, *Beyond the Ivory Tower*, London, 1970

CHAPTER 13: CAN WE RE-POSSESS THE FUTURE?

1 Peter F. Drucker, *The Age of Discontinuity*, New York, London, 1969
2 Filmer Paradise, article 'Putting on the Pressure', in *Drive* magazine, October 1969
3 Richard N. Titmuss, *The Gift Relationship*, London, 1970
4 Claus Moser, 'Measuring the Quality of Life', in *New Society* 10 December 1970. Also 'Social Indicators–Health' by Culyer, Lavers and Williams, in *Social Trends*, 1972, HMSO

5 Ralph Glasser, *The New High Priesthood: ethical and political implications of a marketing orientated society*, London, 1967

6 Alvin Toffler, *Future Shock*, New York, London, 1970

7 Royal Commission on Local Government 1966-9, Minority Report (Cmnd 4040)

8 Correspondence columns of the London *Times*, 22 November 1971

9 *The Lancet* II: 7672, 1970

10 Report of British Medical Association working party, May 1971; Patient Care Index: *The Times Business News*, 13 December 1971

11 Peter Gill, 'Sponsored Sport', *Sunday Telegraph*, 31 October 1971

12 Reports in *Time* magazine, 7 September 1970, 16 August 1971

13 Lord Francis-Williams, *The Right to Know*, London, 1969

14 Louis Lundberg, speech at San Francisco, September 1970

INDEX